OCEAN PLANET

Writings and Images of the Sea

Original text by Peter Benchley

Edited by Judith Gradwohl

Harry N. Abrams, Inc., Publishers,

and Times Mirror Magazines, Inc.,

in association with the Smithsonian Institution

CONTENTS

FOREWORD

It is sometimes difficult for us, as earthbound creatures, to realize that land covers less than a third of the Earth's surface. Beyond the margins of land lie vast oceans that profoundly affect our lives. The seas influence world climate and provide essential resources. They are a major source of protein for billions of people and contribute trillions of dollars to the global economy. They have shaped, sustained, and inspired human cultures for thousands of years.

The oceans' seamless surfaces belie the complex topography and exuberant diversity of life below. Because of their sheer volume, the oceans provide around ninety-nine percent of the living space on the planet. From the miniature bacteria and algae that live in the microlayer — the paper-thin uppermost tier of the ocean — to the strange inhabitants of fissures in the sea floor miles below, every part of the oceans supports some form of life.

Our needs and our growing numbers are creating problems offshore. Symptoms of decline are visible in diminishing catches of fish, in increasing numbers of beach and shellfish-bed closings, in garbage lining the shores of desert isles, in vanishing coral reefs, and in the proliferation of bulkheads and breakwaters along the coasts.

More troubling still is the pollution that falls from the air and pours into the seas from rivers and streams, runs off from urban roads and rural farms, and overflows from sewage treatment plants. Difficult to measure and monitor, pollution can originate hundreds of miles inland, from industry, agriculture, and even individual homes. It is this "nickel and diming" of the oceans through a multitude of small assaults that may ultimately pose the greatest threat.

To commemorate the twenty-fifth anniversary of Earth Day and focus attention on conservation issues affecting watersheds, coasts, and the open sea, the Smithsonian Institution produced the National Forum on Ocean Conservation and Ocean Planet, a traveling exhibition on ocean conservation. From its inception nearly one hundred and fifty years ago, the Smithsonian has supported research to further our understanding of the oceans. Smithsonian scholars study marine science, history, and the interactions between people and the seas. Our own scientists and other specialists worldwide, from a variety of disciplines, contributed information and advice to the Ocean Planet project.

The diversity and complexity of problems threatening the health of oceans compel us to take a wider view of the environment. Successful marine conservation will require precedent-setting forms of local, national, and international cooperation, coupled with the simple realization that our actions on land have far-reaching consequences. Progress is being made, but much work is left to be done. It is our hope that Ocean Planet will bring greater public recognition of the unimaginable grandeur of the oceans, their profound influence on our lives, and the issues that affect the health of the oceans and the quality of life on the planet.

— *I. Michael Heyman, Secretary, Smithsonian Institution*

INTRODUCTION

No matter who we are, no matter where we live or how we make a living, the oceans touch all of our lives. They sustain us, nourish us, inspire us.

Ocean Planet explores the diversity of human connections to the oceans, using writings, photographs, and objects to evoke the experiences of people whose lives have been deeply engaged by the sea. Designed to accompany the Smithsonian Institution's traveling exhibition Ocean Planet, this anthology draws upon themes and images featured in the exhibition.

Twenty selections, ranging from great literature and nature writing to reportage and firsthand accounts, illuminate the perceptions and beliefs that forge personal bonds to the sea and show how committed writers have come to grips with the changing relationship between humanity and the oceans. The stories they tell are both joyous and tragic. Portfolios of photographs and collections of significant facts amplify the ideas expressed in the selections. The book as a whole explores four major themes: perceptions of the seas, the lives of seafarers, scientific discovery, and environmental threats to the seas.

In "Visions of the Sea" we celebrate the oceans as a source of inspiration, and we examine the many prisms through which humankind sees the sea. There is no consensus, save that the oceans command respect. In these writings, the sea can be cruel and magnificent, patient and relentless, romantic and scientific, savage and serene.

"Seafarers" focuses on people who go to sea, particularly sailors and fishermen. Despite their wide diversity, most share a strong sense of community. They develop and pass along their knowledge of the lore and the ways of the sea, and they face a life of risk and uncertainty.

The difficulties of exploring the sea's extreme pressure, inky darkness, and icy temperatures have severely limited our knowledge of vast areas of the world's oceans. In "Discovery" we present personal perspectives about ocean science and exploration, to help us understand why people plumb the ocean depths and to let us share the thrill of scientific discovery.

The prodigious size and great age of the oceans are almost incomprehensible. Their magnificent power far outstrips any technology we could develop to tame them. They seem omnipotent and infinite. Yet, they are vulnerable. The oceans and the life they support are showing signs of human interference. "Oceans in Peril" focuses on environmental threats to the delicate balance between land and sea, and on possible solutions.

Ultimately, *Ocean Planet* is about mutual dependency. We must cherish the oceans and treat them with care for our own lives and for the future of life on Earth.

— *Judith Gradwohl and Peter Benchley*

VISIONS OF THE SEA

VISIONS OF THE SEA

I have a favorite place. I've been there only twice, but I revisit it often, whenever I feel rattled or suffocated by the press of daily life. I go there in my mind, and find serenity.

It is blue, this place of mine. There are no reds or yellows, and only a hint of green, though sunlight from above casts shafts that look like gold. Fish swim by on silent patrol, unhurried, unconcerned with me, as if sensing that I pose no threat, that my quest is for peace.

My place is a sand shelf, sixty feet beneath the surface of the sea, surrounding the Turks and Caicos Islands, south of the Bahamas. It is a tiny patch of flatness at the base of a gentle slope, a plateau that ends abruptly in a sheer drop into the abyss. Sometimes I swim to the edge and look down into the deepening blue that darkens finally into blackness, into a world I will never know, where wondrous creatures respond to the natural rhythms of life and death beyond the reach of man.

After half an hour or so, I surface, and I feel revived, resuscitated. For I take nourishment from the sea, in reality or imagination. It speaks to me of continuity, of promise, of adventure.

Some people hate it, or fear it, or feel threatened by it. Being on the water gives them an uneasy feeling of exposure and vulnerability; being under the water makes them claustrophobic.

Not me. Even in times of danger, I don't personify the sea as a foe. I try not to personify it at all. I don't think of it as angry or vengeful, vicious or merciless. It is what it is, a force of nature — in company with the sun and the wind — possessed of inconceivable power that is sometimes destructive but more often, far more often, benevolent, vital to the sustenance of life on the planet.

Throughout history, our visions of the sea have been as varied as the moods we ascribe to the sea itself. Ambivalence has been the only constant in the attitudes of human beings toward the ocean environment that once was womb to the entire race.

In the selections that follow, Kenneth Grahame portrays the ocean as a highway to romantic adventure, exulting in "the sounding slap of the great green seas," while, writing at more or less the same time, Joseph Conrad declares it to be "the irreconcilable enemy of ships and men ever since ships and men had the unheard-of audacity to go afloat together in the face of his frown."

It has always struck me as fascinating that Conrad, who spent much of his life at sea, and devoted much of his prose to the sea, seemed often to loathe the sea with an abiding passion. "The most amazing wonder of the deep," he wrote, "is its unfathomable cruelty."

The only time I recall anthropomorphizing the sea, ascribing human characteristics to it, was on a three-day trip across the Gulf of Mexico in a small boat, in a terrible storm, with our compass smashed and all other navigational gear out of order, with no idea where we were or which way we were headed, and with a yawing wind that caused a confused, unsteady sea.

I was afraid, of course, though I harbored a confidence — foolish and unfounded — that even if we capsized or sank I would survive by surrendering to the sea, not fighting it, by floating with my back to the cresting waves.

I wasn't resentful, I didn't blame the sea. All I remember thinking was, "She's putting us through our paces this time." Conrad's ocean, you see, is personified, in the Classical manner, as a masculine deity; mine is a goddess.

Until very recently, the sea was seen as immortal, invulnerable, eternal. In 1818, Lord Byron wrote confidently:

Roll on, thou deep and dark blue ocean — roll!
Ten thousand fleets sweep over thee in vain;
Man marks the earth with ruin — his control
Stops with the shore.

Even Rachel Carson, who was one of the first major writers to sound environmental alarums, perceived the sea as eternal and used it as a metaphor for life, which she described as "a force strong and purposeful, as incapable of being crushed or diverted from its ends as the rising tide."

Today, her confidence seems misplaced, naive. For though we have long since learned that we cannot tame the sea, we are just now becoming aware that it *is* in our power to destroy the sea. Man's effect, his scourge, extends far beyond the shore, beyond the continental shelves, into the deepest recesses of abyssal canyons.

As recently as 1969, in "The Star Thrower," the mystical naturalist Loren Eiseley made up a parable of salvation by opposing human will and nature's chaos, a vision in which the sea symbolizes death, disgorging its helpless creatures onto the sand with every wave: "The beaches of Costabel are littered with the debris of life."

Today, we are more likely to remark that the beaches are littered with . . . litter. And what writer would now have the confidence to celebrate our humanity in the face of nature?

The tide, we now know, may not always rise and fall with healthy cadence, unless *we* change our ways. We may love the sea or hate it; we may seek it out or fear it. But we all share one absolute imperative: we must treat the sea with respect. We cannot continue to use it as a dumping ground; we cannot continue to poison it; we cannot strip from it the life it produces and return to it only the waste we produce.

We do not deserve, nor can we afford, to treat anything in nature — on land or sea — with contempt. To do so betrays self-destructive ignorance.

As John Wiley discovers on his beach walk with physicist Jim Trefil, even the most mundane, rhythmic repetitions of the sea reflect "the basic forces that run the Universe" and contribute to keeping us all alive.

No one has ever said it better than Henry Beston, in his 1928 classic, *The Outermost House.* He is writing about animals only here, but, like all fine writing, it is larger in implication.

"We need another and a wiser and perhaps a more mystical concept of animals. Remote from universal nature, and living by complicated artifice, man in civilization surveys the creature through the glass of his knowledge and sees thereby a feather magnified and the whole image in distortion. We patronize them for their incompleteness, for their tragic fate of having taken form so far below ourselves. And therein we err, and greatly err. For the animal shall not be measured by man. In a world older and more complete than ours they move finished and complete, gifted with extensions of the senses we have lost or never attained, living by voices we shall never hear. They are not brethren, they are not underlings; they are other nations, caught with ourselves in the net of life and time, fellow prisoners of the splendour and travail of the earth."

The sea is not eternal.

Let us not be the ones to ruin it.

— PETER BENCHLEY

WAYFARERS ALL

FOR MANY THE OPEN OCEAN SYMBOLIZES ADVENTURE AND FREEDOM — FROM BOREDOM AND ROUTINE, FROM SOCIETY'S CONSTRAINTS, FROM PERSONAL RESPONSIBILITIES, FROM POLITICAL OPPRESSION. IN THIS ROMANTIC EPISODE FROM KENNETH GRAHAME'S CLASSIC CHILDREN'S STORY, *THE WIND IN THE WILLOWS*, A SEA RAT ENCOUNTERS A WATER RAT IN AN ENGLISH COUNTRY LANE AND WEAVES A HYPNOTIC TALE OF A MEDITERRANEAN VOYAGE. AS THE MERRY SEA RAT TRAVELS A WATERY HIGHWAY BETWEEN FAMILIAR DESTINATIONS, HE IS AT HOME IN EVERY LAND AND SUBJECT TO NONE.

First-class passengers settle in on the deck of the Cunard Line's Saxonia *for the run from New York to Liverpool on a spring day in 1924.*

M y last voyage," began the Sea Rat, "that landed me eventually in this country, bound with high hopes for my inland farm, will serve as a good example of any of them, and, indeed, as an epitome of my highly-coloured life. Family troubles, as usual, began it. The domestic storm-cone was hoisted, and I shipped myself on board a small trading vessel bound from Constantinople, by classic seas whose every wave throbs with a deathless memory, to the Grecian Islands and the Levant. Those were golden days and balmy nights! In and out of harbour all the time — old friends everywhere — sleeping in some cool temple or ruined cistern during the heat of the day — feasting and song after sundown, under great stars set in a velvet sky! Thence we turned and coasted up the Adriatic, its shores swimming in an atmosphere of amber, rose, and aquamarine; we lay in wide land-locked harbours, we roamed through ancient and noble cities, until at last one morning, as the sun rose royally behind us, we rode into Venice down a path of gold. O, Venice is a fine city, wherein a rat can wander at his ease and take his pleasure! Or, when weary of wandering, can sit at the edge of the Grand Canal at night, feasting with his friends, when the air is full of music and the sky full of stars, and the lights flash and shimmer on the polished steel prows of the swaying gondolas, packed so that you could walk across the canal on them from side to side! And then the food — do you like shellfish? Well, well, we won't linger over that now."

He was silent for a time; and the Water Rat, silent too and enthralled, floated on dream-canals and heard a phantom song pealing high between vaporous grey wave-lapped walls.

"Southwards we sailed again at last," continued the Sea Rat, "coasting down the Italian shore, till finally we made Palermo, and there I quitted for a long, happy spell on shore. I never stick too long to one ship; one gets narrow-minded and prejudiced. Besides, Sicily is one of my happy hunting-grounds. I know everybody there, and their ways just suit me. I spent many jolly weeks in the island, staying with friends up country. When I grew restless again I took advantage of a ship that was trading to Sardinia and Corsica; and very glad I was to feel the fresh breeze and the sea-spray in my face once more."

"But isn't it very hot and stuffy, down in the — hold, I think you call it?" asked the Water Rat.

The seafarer looked at him with the suspicion of a wink. "I'm an old hand," he remarked with much simplicity. "The captain's cabin's good enough for me."

"It's a hard life, by all accounts," murmured the Rat, sunk deep in thought.

"For the crew it is," replied the seafarer gravely, again with the ghost of a wink.

"From Corsica," he went on, "I made use of a ship that was taking wine to the mainland. We made Alassio in the evening, lay to, hauled up our wine-casks, and hove them overboard, tied one to the other by a long line. Then the crew took to the boats and rowed shorewards, singing as they went, and drawing after them the long bobbing procession of casks, like a mile of porpoises. On the sands they had horses waiting, which dragged the casks up the steep street of the little town with a fine rush and clatter and scramble. When the last cask was in, we went and refreshed and rested, and sat late into the night, drinking with our friends; and next morning I was off to the great olive-woods for a spell and a rest. For now I had done with islands for the time, and ports and shipping were plentiful; so I led a lazy life among the peasants, lying and watching them work, or stretched high on the hillside with the blue Mediterranean far below me. And so at length, by easy stages, and partly on foot, partly by sea, to Marseilles, and the meeting of old shipmates, and the visiting of great ocean-bound vessels, and feasting once more. Talk of shell-fish! Why, sometimes I dream of the shell-fish of Marseilles, and wake up crying!" . . .

The Sea Rat, as soon as his hunger was somewhat assuaged, continued the history of his latest voyage, conducting his simple hearer from port to port of Spain, landing him at Lisbon, Oporto, and Bordeaux, introducing him to the pleasant harbours of Cornwall and Devon, and so up the Channel to that final quayside, where, landing after winds long contrary, storm-driven and weather-beaten, he had caught the first magical hints and heraldings of another Spring, and, fired by these, had sped on a long tramp inland, hungry for the experiment of life on some quiet farmstead, very far from the weary beating of any sea.

Spellbound and quivering with excitement, the Water Rat followed the Adventurer league by league, over stormy bays, through crowded roadsteads, across harbour bars on a racing tide, up winding rivers that hid their busy little towns round a sudden turn, and left him with a regretful sigh planted at his dull inland farm, about which he desired to hear nothing.

By this time their meal was over, and the Seafarer, refreshed and strengthened, his voice more vibrant, his eye lit with a brightness that seemed caught from some far-away sea-beacon, filled his glass with the red and glowing vintage of the South, and, leaning towards the Water Rat, compelled his gaze and held him, body and soul, while he talked. Those eyes were of the changing foam-streaked grey-green of leaping Northern seas; in the glass shone a hot ruby that seemed the very heart of the South, beating for him who had courage to respond to its pulsation. The twin lights, the shifting grey and the steadfast red, mastered the Water Rat and held him bound, fascinated, powerless. The quiet world outside their rays receded far away and ceased to be. And the talk, the wonderful talk flowed on — or was it speech entirely, or did it pass at times into song — chanty of the sailors weighing the dripping anchor, sonorous hum of the shrouds in a tearing North-Easter, ballad of the fisherman hauling his nets at sundown against an apricot sky, chords of guitar and mandoline from gondola or caique? Did it change into the cry of the wind, plaintive at first, angrily shrill as it freshened, rising to a tearing whistle, sinking to a musical trickle of air from the leech of the bellying sail? All these sounds the spellbound listener seemed to hear, and with them the hungry complaint of the gulls and the sea-mews, the soft thunder of the breaking wave, the cry of the protesting shingle. Back into speech again it passed, and with beating heart he was following the adventures of a dozen seaports, the fights, the escapes, the rallies, the comradeships, the gallant undertakings; or he searched islands for treasure, fished in still lagoons and dozed day-long on warm white sand. Of deep-sea fishings he heard tell, and mighty silver gatherings of the mile-long net; of sudden perils, noise of breakers on a moonless night, or the tall bows of the great liner taking shape overhead through the fog; of the merry home-coming, the headland rounded, the harbour lights opened out:

the groups seen dimly on the quay, the cheery hail, the splash of the hawser; the trudge up the steep little street towards the comforting glow of red-curtained windows.

Lastly, in his waking dream it seemed to him that the Adventurer had risen to his feet, but was still speaking, still holding him fast with his sea-grey eyes.

"And now," he was softly saying, "I take to the road again, holding on south-westwards for many a long and dusty day; till at last I reach the little grey sea town I know so well, that clings along one steep side of the harbour. There through dark doorways you look down flights of stone steps, overhung by great pink tufts of valerian and ending in a patch of sparkling blue water. The little boats that lie tethered to the rings and stanchions of the old sea-wall are gaily painted as those I clambered in and out of in my own childhood; the salmon leap on the flood tide, schools of mackerel flash and play past quaysides and foreshores, and by the windows the great vessels glide, night and day, up to their moorings or forth to the open sea. There, sooner or later, the ships of all seafaring nations arrive; and there, at its destined hour, the ship of my choice will let go its anchor. I shall take my time, I shall tarry and bide, till at last the right one lies waiting for me, warped out into midstream, loaded low, her bowsprit pointing down harbour.

I shall slip on board, by boat or along hawser; and then one morning I shall wake to the song and tramp of the sailors, the clink of the capstan, and the rattle of the anchor-chain coming merrily in. We shall break out the jib and the foresail, the white houses on the harbour side will glide slowly past us as she gathers steering-way, and the voyage will have begun! As she forges towards the headland she will clothe herself with canvas; and then, once outside, the sounding slap of great green seas as she heels to the wind, pointing South!

"And you, you will come too, young brother; for the days pass, and never return, and the South still waits for you. Take the Adventure, heed the call, now ere the irrevocable moment passes! 'Tis but a banging of the door behind you, a blithesome step forward, and you are out of the old life and into the new! Then some day, some day long hence, jog home here if you will, when the cup has been drained and the play has been played, and sit down by your quiet river with a store of goodly memories for company. You can easily overtake me on the road, for you are young, and I am ageing and go softly. I will linger, and look back; and at last I will surely see you coming, eager and light-hearted, with all the South in your face!"

— KENNETH GRAHAME,
The Wind in the Willows, 1908

Sanderlings, migratory shore birds, run along the water's edge, feeding in the area exposed by retreating waves.

several hundred million tons of dried organic matter will find their way into the atmosphere in this way.

Back in the parking lot, tired and sunburned, my mind is whirling with the sensations of having seen and understood elemental forces at work. Jim has been taken with the little I've been able to tell him about birds and the people who study them, but neither of us shows signs of switching allegiance. "It's beautiful and interesting," he offers. "But it just isn't physics."

— JOHN P. WILEY, JR.,
Smithsonian Magazine, July 1985

SEAFARERS

SEAFARERS

On a rainy afternoon not long ago, I was wandering through a cemetery in a seaport town. The town had been born in the 1600s, had thrived as a center of the whaling industry until the middle of the 1800s, and had by now settled into a peaceful old age as a village sustained by summer residents and tourists.

The history of the town was etched on the cemetery's headstones. One after another, they spoke with stark simplicity of the fate of many of the townspeople.

"Lost At Sea."

"Swept Away."

"Drowned."

One stone in particular caught my eye. It memorialized a man who had died in 1846 at the age of twenty-four. No tender poem graced the granite, no biblical exhortation. Beneath the notation "Perished At Sea" were two words that, to me, spoke volumes:

"Welcome Home."

What struck me about those words was that they bespoke no anger, no resentment; they did not rail against death or nature or injustice.

They were words of acceptance, acceptance not only of the inevitability of death but of its rightness. To me they said, in effect, "From the sea we come, on the sea we live, to the sea we must return."

But at *twenty-four?*

What kind of person chooses a life in which an early death is not only a risk but a probability? What leads a man or woman to leave the land and venture into a world alien and unknown — cold, fickle, forbidding, and possessed of a power beyond anything the mind can imagine?

Of course, the sea should not be alien to us, nor we to it. Biologically, we are all creatures of the sea. For the first nine months of our existence we are obligate water-breathers, and in the course of our development we pass through stages that confirm our past as animals more at home in the sea than on land.

Yet, we *are* strangers in the ocean. Although three-quarters of our planet is covered by water, most of it is more than two miles deep, so even those of us who spend our lives in, on, and around the sea are able barely to scratch its surface.

Still we persist, many of us. There is a worldwide community of people whose lives are wed to the sea. They are almost a race apart, though they are, in fact, all races, all colors, all creeds. They are not a single society, though they share skills and attitudes and philosophies of life.

And of death.

Fatalism is almost a prerequisite of the seafaring life. "What the sea wants," goes an old British expression, "the sea will have." And, evidently, the sea wants a lot.

PREVIOUS SPREAD
Fishermen use hand-hewn oars to maneuver their rowboat near a salmon set net, Sakhalin Islands in the Sea of Okhotsk, Russia, 1990. The net is hauled in by hand to a larger fishing boat, which takes the catch to a processing plant.

36

Consider: in the United States, which is among the world's most advanced nations in terms of technology and safety, commercial fishing is the single most dangerous profession. The death rate for commercial fishermen is 155 per 100,000 workers, compared to 24 per 100,000 for police officers, 26 per 100,000 for miners, and 50 per 100,000 for taxi drivers.

A list on a wall in the city hall of Gloucester, Massachusetts, records the names of some 4,000 local fishermen who have died at sea since 1874. An estimated 6,000 others perished in the 250 years prior to 1874, and the surname of nearly every fishing family in the town can be found somewhere on the wall.

Men and women go to sea; men and women die. As the captain in John McPhee's *Looking For A Ship* says, "Almost every hour of every day someone is getting it. Right now someone is getting it."

Fatalism, however, is not synonymous with resignation, at least not any more. It used to be that seafarers would not bother to learn how to swim, the theory being that if you went overboard you were as good as dead. Lifejackets were often scorned, for much the same reason. Nowadays, sailors and fishermen in most countries are taught to handle themselves as well in the water as upon it, and boats that don't carry safety and survival gear cannot legally leave port.

Since prehistory, people have employed protective devices — however ephemeral they may have appeared — to give them an edge against the spirits of the sea.

Vessels from Christian lands carry shrines to the Virgin Mary or to any of several saints; sailors wear medals or lucky charms.

Micronesians have placed ornaments in the bows of their canoes — they may be no more than decorated planks of wood — to ward off weather, guide them across the trackless ocean, and defend them against their enemies.

Inuit hunters have sewn amulets in their clothing and protect precious charms with waterproof pouches.

There are things *not* to do, as well, to avoid failure or calamity.

Don't carry an umbrella aboard a boat; don't change the name of a vessel; don't open a hatch while at sea.

In Scotland and Ireland, don't wear clothes dyed with colors made from sea plants, for the sea will want to reclaim them.

In Newfoundland, don't keep the first fish of the day. Spit on it and throw it back, and you will be assured of good fishing.

I remember being in the Turks and Caicos Islands years ago, and finding a tiny eighteenth-century figurine amid some shipwreck debris. I wanted to bring it home, but our captain's wife, a Bermudian, insisted that I throw it overboard before we set sail. "It sank one ship," she said, "and I won't be party to its sinking another."

It may be tempting to dismiss these sensibilities as superstitious quirks, but to do so would be ignorant and foolhardy. Superstitions, like clichés, are usually grounded in basic truths or hard-won experience.

What is a seafarer? Can people from nearly every culture on the globe be bundled together by their traits or eccentricities or attitudes? No, and yes.

They have no secret handshakes; there are no universal shibboleths. But there are special skills, wonderful arcana, and languages peculiar to the sea. In English alone, countless words and phrases hatched at sea have migrated ashore into the landlubber's lexicon: "aloof," "bear down," "scuttlebutt," "skedaddle," "junk," "slush fund," "son of a gun" and "letting the cat out of the bag," to cite but a handful.

When I was a child, I hung around the docks on Nantucket, marvelling at the impossible knots tied by grizzled men with hands as hard as hammers. The knots had names redolent of far-off places and exotic practices: sheepshank, clove hitch, true lover's, Turk's head.

As I grew, and spent more time on the steel-gray waters off the Atlantic Coast and the turquoise shoals of the Great Barrier Reef and the gunmetal blue of the Bermuda deep, I came to believe that there was an element of the magician in every seafarer.

They could predict the weather by the haze in the sky or by a change in the smell of the salt air.

They could identify a school of fish by the way the water roiled.

They could locate seldom-visited spots in the open ocean, not by landmarks — for there were none — but by the contours of the bottom and the patterns of the tidal runs.

Seafarers are possessed of many other qualities as well: an ability to endure boredom, loneliness, and separation; instincts, honed by experience, that trigger instantaneous responses to sudden emergencies; loyalty to one another, coupled with faith in themselves; a fierce independence, sometimes at the expense of comfort and family. As Milt Miller says in Peter Matthiessen's *Men's Lives,* "Independence costs you a lot of money."

And this above all is possessed by every seafarer: a profound, abiding, unwavering respect for the sea. He or she may enjoy the sea, may profit from it, may hate it and wish to leave it, but, one and all, they know that their lives continue only at the sufferance of the sea.

No wonder, then, that they personify it as a god or demon, a friend or enemy. For, come the day when it turns upon them. . . .

Welcome home.

— PETER BENCHLEY

SINGLEHANDER

WHEN ANN DAVISON CROSSED THE ATLANTIC FROM PLYMOUTH, ENGLAND, TO ANTIGUA, IN THE WEST INDIES, IN 1952, SHE BECAME THE FIRST WOMAN EVER TO SAIL ACROSS AN OCEAN ALONE. HER ACCOMPLISHMENT WAS MADE EVEN MORE EXTRAORDINARY BY THE SIZE OF HER BOAT *FELICITY ANN* — 23 FEET LONG, 7 1/2 FEET WIDE — AND BY THE FACT THAT A PREVIOUS ATTEMPT, IN A 70-FOOT KETCH, HAD ENDED IN A SHIPWRECK THAT HAD CLAIMED THE LIFE OF HER HUSBAND. HER JOURNAL OF THE LEG FROM CASABLANCA TO THE CANARY ISLANDS CAPTURES MANY OF THE ELEMENTS OF SOLO SAILING: THE COURAGE, THE ENDURANCE, THE LONELINESS, THE PEACE, THE SERENITY, THE NEED FOR CONSTANT VIGILANCE, THE SENSE OF CONNECTION WITH THE LIVING SEA. AND ALWAYS, JUST OVER THE HORIZON, THE UNKNOWN DANGERS AND THE UNEXPECTED PLEASURES — NOT ONLY WIND AND WEATHER BUT ALSO PASSING SHIPS.

View off the stern of the Duracell *of deep swells in the Southern Ocean. The 60-foot boat was competing in the Vendee Globe Challenge, a race around the world, 1990.*

Conditions had a delicious dreamy Southern feel about them, calm and unhurried. There were lovely soft pearl-grey nights of a peculiar luminosity and soothing restfulness that were the physical manifestation of contentment. There were sunrises of such crystalline clarity and pristine glory that one could forgive any amount of travail for the joy of beholding those few golden moments when the world was born anew. There were sunsets so lurid, when an orange sun crept down a black and blood-red sky into a smooth lead-coloured sea, that one was convinced there was nothing less than a hurricane in the offing. I would shorten sail and batten down and prepare for the worst, only to discover that all the fuss in the heavens was for a few drops of rain. The weather eye I had acquired through years of flying and farming in England was sadly out in the lower latitudes, where the familiar signs and portents meant nothing at all. The weather could, and did, change with extraordinary rapidity, and the minutest rise or fall in barometric pressure might mean a severe blow, or nothing. . . . I soon gave up trying to forecast and took the weather as it came. After all, there is very little else you can do in the ocean, with no convenient ports to run to for shelter there, so I gave up reefing until it was necessary, and it was hardly ever necessary on this trip, as most of the time there was either a glass calm or a very light breeze, and our average day's progress was twenty miles.

The snail-like advance was a straight invitation to barnacles to grow on the log line, and they were surprisingly tenacious and difficult to remove. The water was so still and clear that sometimes it was almost as if you could see straight down to the bottom of the sea. Fascinating little striped fish, black and bright blue, swam about in the shade of the ship. A few flying fish skittered across the surface like flat stones thrown on a pond. They were very small flying fish, no bigger than minnows. There were times when rubbish thrown over the side in the morning would still be alongside at nightfall. Then the air was breathless and there would not be the smallest sound from the ship, not even a creak, and the silence was primeval. One might have been alone on the planet where even a cloud spelt companionship.

Most of the time, however, there was a huge swell in which *FA* rolled abominably and flung her boom from side to side with a viciousness that threatened to wrench it clean out of its fastenings. She rattled her blocks and everything not immovably fast below with an aggravating irregularity, so that I was driven to a frenzy of restowing and rigging preventers in an effort to restore peace. An intermittent blop — rattle — crash on a small boat at sea is the nautical version of the Chinese water torture.

Calms permit a little basking, but not much for a single-handed sailor. They provide an opportunity to overhaul gear and repair or renew anything that might give way under more embarrass-

Women compete to defend Americas Cup for the San Diego Yacht Club in May 1995. The team, fielded by Bill Koch's America3, is seen here on one of its two training boats.

ing circumstances, for if there is one thing the sea will not forgive it is a lost opportunity. I made up and reeved new jib sheets, mended slide-seizings on the mainsail, patched the sails where they showed signs of chafe, and recovered the fenders whose canvas covers had been ruined by oil in the dock at Gibraltar, and felt no end salty at my work, deriving a deep satisfaction in the doing of it, even though the patches on the sails were by no means the finest examples of a sailmaker's art.

For the first nine days out of Casablanca there was not a ship to be seen, and I missed them, grizzling quietly to myself at the loneliness; then we joined the north- and south-bound shipping lane and two steamers appeared on the horizon at the same time, whereon, embarrassed by riches perhaps, I perversely resented their presence. "What are you doing on my ocean?"

Being in the shipping lane again meant the resumption of restless, sleepless nights. I figured

out it took twenty minutes for a ship invisible over the horizon to reach us, and as a big ship was extremely unlikely to see me I had to see her, so any rest below was broken every twenty minutes throughout the hours of darkness. Enough practice since leaving England had endowed me with a personal alarm system which rang me out of a comatose condition at the appropriate intervals. Occasionally it let me oversleep, and once I awoke to find a south-bound steamer twenty-five yards astern of us. . . . A miss is as good as a mile maybe, but twenty-five yards is a narrow enough margin in the ocean, and it gave the required jolt to the personal alarm clock. On these ship-watching nights I used to get two hours of genuine sleep at dawn, when it could be assumed that *FA* was reasonably visible, and I couldn't care less by then anyway, but the overall lack of sleep did not improve the general physical condition, already much lowered by dysentery. The thought processes, never on Einstein levels, were reduced to a positively moronic grasp, and I had some rare hassles with navigational problems. However, the balance of nature was somewhat restored in that I was eating better on this trip than on any of the previous ones — the voyage from Douarnenez to Vigo was made almost exclusively on oranges — and there are several references to cooked meals in the log book. . . . I had an uncomplicated yearning for plain boiled potatoes and cabbage. As these do not represent a normal taste on my part, I concluded it was a deficiency desire, and stepped up the daily dose of vitamin tablets: a strict necessity for ocean voyagers, as I discovered on the nineteen-day Vigo to Gibraltar run, when I tried to do without them and broke out into reluctant-to-heal sores. The only canned goods whose vitamin content survives the canning process are tomatoes, which probably explains why canned foods lost all appeal for me as soon as I went to sea. Very practically I was learning what stores would be required for the long passage.

One supper was especially memorable, though not for the menu. At 1750 hours, Sunday, October 5th to be exact, I was fixing some cheese nonsense on the stove, for it was a flat calm and I was in an experimental mood, and whilst stirring

the goo in the pan I happened to glance through the porthole over the galley and spied a steamer way over on the horizon, the merest speck to eastward of us, going south. A few minutes later I looked out again and to my surprise saw she had altered course and was making towards us. Coming out of her way specially to look at a little ship. Thrilled to the quick, I abandoned supper, brushed my hair, and made up my face, noting with detached amazement that my hands were trembling and my heart was beating, and I was as excited as if I was preparing for a longed-for assignation.

She was a tall, white-grey Italian liner, the *Genale* of Rome, and she swept round astern of us, the officers on her bridge inspecting *FA* keenly through their binoculars. As she had so kindly come many miles out of her way, I had no wish to delay her needlessly, for minutes are valuable to a ship on schedule, so I made no signals, but waved, and the whole ship seemed to come alive with upraised arms waving in reply. She went on her way satisfied that all was well with her midget counterpart, and the night was a little less lonely from the knowledge of her consideration.

— ANN DAVISON, *My Ship Is So Small*, 1956

WOMEN'S WORK

ALTHOUGH DEEP-SEA FISHING IS USUALLY
A MAN'S OCCUPATION, WOMEN ALSO GO
TO SEA. WOMEN HAVE ALWAYS PLAYED
IMPORTANT ROLES IN FISHERIES AS MAN-
AGERS, MARKETERS, AND PROCESSORS.

JUANITTA LANG, SKIPPER
OF THE *VAGABOND*, A
38-FOOT WOODEN TROLLER
THAT FISHES SOUTHEAST-
ERN ALASKAN WATERS.
SHE IS HOLDING A SALMON.

O V E R L E A F WOMEN
FISHING IN ÎLE AUX
NATTES, MADAGASCAR.
WOMEN THROUGHOUT
THE WORLD HAVE
TRADITIONALLY FISHED
NEAR SHORE.

SELLING THE FRESH
CATCH IN AN OPEN-AIR
MARKET, DAKAR,
SENEGAL.

THE SEA HAS NO GENEROSITY

THE SEA IS TERRIBLE AND TERRIFYING. SAILORS AND FISHERMEN THROUGHOUT THE AGES HAVE TRIED, IN THEIR MINDS, TO TAME IT, TO UNDERSTAND IT, TO BEFRIEND IT. NOT WRITER JOSEPH CONRAD, WHO WENT TO SEA AT AGE SEVENTEEN AND WAS CERTIFIED AS A MASTER AT TWENTY-SEVEN. IN THIS PASSAGE, HE NARRATES THE EXPERIENCE THAT STRIPPED HIM FOREVER OF HIS ROMANTIC ILLUSION THAT THE SEA MIGHT NURTURE HUMAN PLANS AND HOPES.

World War II saw many dramatic ocean rescues. Lieutenant Commander J. T. Blackburn spent sixty-four hours in the Atlantic. Here he is being hoisted from a tanker to the USS Senate, November 1942.

The sea — this truth may be confessed — has no generosity. No display of manly qualities — courage, hardihood, endurance, faithfulness — has ever been known to touch its irresponsible consciousness of power. The ocean has the conscienceless temper of a savage autocrat spoiled by much adulation. He cannot brook the slightest appearance of defiance, and has remained the irreconcilable enemy of ships and men ever since ships and men had the unheard-of audacity to go afloat together in the face of his frown. From that day he has gone on swallowing up fleets and men without his resentment being glutted by the number of victims — by so many wrecked ships and wrecked lives. Today, as ever, he is ready to beguile and betray, to smash and to drown the incorrigible optimism of men who, backed by the fidelity of ships, are trying to wrest from him the fortune of their house, the dominion of their world, or only a dole of food for their hunger. If not always in the hot mood to smash, he is always stealthily ready for a drowning. The most amazing wonder of the deep is its unfathomable cruelty.

I felt its dread for the first time in mid-Atlantic one day, many years ago, when we took off the crew of a Danish brig homeward bound from the West Indies. A thin, silvery mist softened the calm and majestic splendour of light without shadows — seemed to render the sky less remote and the ocean less immense. It was one of the days, when the might of the sea appears indeed lovable, like the nature of a strong man in moments of quiet intimacy. At sunrise we had made out a black speck to the westward, apparently suspended high up in the void behind a stirring, shimmering veil of silvery blue gauze that seemed at times to stir and float in the breeze which fanned us slowly along. The peace of that enchanting forenoon was so profound, so untroubled, that it seemed that every word pronounced loudly on our deck would penetrate to the very heart of that infinite mystery born from the conjunction of water and sky. We did not raise our voices. "A water-logged derelict, I think, sir," said the second officer quietly, coming down from aloft with the binoculars in their case slung across his shoulders; and our captain, without a word, signed to the helmsman to steer for the black speck. Presently we made out a low, jagged stump sticking up forward — all that remained of her departed masts.

The captain was expatiating in a low conversational tone to the chief mate upon the danger of these derelicts, and upon his dread of coming upon them at night, when suddenly a man forward screamed out, "There's people on board of her, sir! I see them!" in a most extraordinary voice — a voice never heard before in our ship; the amazing voice of a stranger. It gave the signal for a sudden tumult of shouts. The watch below ran up the forecastle head in a body, the cook dashed out of the galley. Everybody saw the poor fellows now. They were there! And all at once our

*Wreck of a Japanese fishing
vessel on a reef off Kure Atoll,
Hawaii, 1970s.*

ship, which had the well-earned name of being
without a rival for speed in light winds, seemed
to us to have lost the power of motion, as if the
sea, becoming viscous, had clung to her sides. And
yet she moved. Immensity, the inseparable companion of a ship's life, chose that day to breathe
upon her as gently as a sleeping child. The clam-
our of our excitement had died out, and our living
ship, famous for never losing steerage way as long
as there was air enough to float a feather, stole,
without a ripple, silent and white as a ghost,
towards her mutilated and wounded sister, come
upon at the point of death in the sunlit haze of a
calm day at sea.

With the binoculars glued to his eyes, the
captain said in a quavering tone: "They are waving

to us with something aft there." He put down
the glasses on the skylight brusquely, and began
to walk about the poop. "A shirt or a flag," he
ejaculated irritably. "Can't make it out. . . . Some
damn rag or other!" He took a few more turns
on the poop, glancing down over the rail now and
then to see how fast we were moving. His nervous
footsteps rang sharply in the quiet of the ship,
where the other men, all looking the same way,
had forgotten themselves in a staring immobility.
"This will never do!" he cried out suddenly.
"Lower the boats at once! Down with them!"

Before I jumped into mine he took me aside,
as being an inexperienced junior, for a word of
warning:

"You look out as you come alongside that

she doesn't take you down with her. You understand?"

He murmured this confidentially, so that none of the men at the falls should overhear, and I was shocked. "Heavens! as if in such an emergency one stopped to think of danger!" I exclaimed to myself mentally, in scorn of such cold-blooded caution.

It takes many lessons to make a real seaman, and I got my rebuke at once. My experienced commander seemed in one searching glance to read my thoughts on my ingenuous face.

"What you're going for is to save life, not to drown your boat's crew for nothing," he growled severely in my ear. But as we shoved off he leaned over and cried out: "It all rests on the power of your arms, men. Give way for life!"

We made a race of it, and I would never have believed that a common boat's crew of a merchantman could keep up so much determined fierceness in the regular swing of their stroke. What our captain had clearly perceived before we left had become plain to all of us since. The issue of our enterprise hung on a hair above that abyss of waters which will not give up its dead till the Day of Judgment. It was a race of two ship's boats matched against Death for a prize of nine men's lives, and Death had a long start. We saw the crew of the brig from afar working at the pumps — still pumping on that wreck, which already had settled so far down that the gentle, low swell, over which our boats rose and fell easily without a check to their speed, welling up almost level with her head-rails, plucked at the ends of broken gear swinging desolately under her naked bowsprit.

We could not, in all conscience, have picked out a better day for our regatta had we had the free choice of all the days that ever dawned upon the lonely struggles and solitary agonies of ships since the Norse rovers first steered to the westward against the run of Atlantic waves. It was a very good race. At the finish there was not an oar's length between the first and second boat, with Death coming in a good third on the top of the very next smooth swell, for all one knew to the contrary. The scuppers of the brig gurgled softly all together when the water rising against her sides subsided sleepily with a low wash, as if playing about an immovable rock. Her bulwarks were gone fore and aft, and one saw her bare deck low-lying like a raft and swept clean of boats, spars, houses — of everything except the ringbolts and the heads of the pumps. I had one dismal glimpse of it as I braced myself to receive upon my breast the last man to leave her, the captain, who literally let himself fall into my arms.

It had been a weirdly silent rescue — a rescue without a hail, without a single uttered word, without a gesture or a sign, without a conscious exchange of glances. Up to the very last moment those on board stuck to their pumps, which spouted two clear streams of water upon their bare feet. Their brown skin showed through the rents of their shirts; and the two small bunches of half-naked, tattered men went on bowing from the waist to each other in their back-breaking labour, up and down, absorbed, with no time for a glance over the shoulder at the help that was coming to them. As we dashed, unregarded, alongside, a voice let out one, only one hoarse howl of command, and then, just as they stood, without caps, with the salt drying grey in the wrinkles and folds of their hairy, haggard faces, blinking stupidly at us their red eyelids, they made a bolt away from the handles, tottering and jostling against each other, and positively flung themselves over upon our very heads. The clatter they made tumbling into the boats had an extraordinarily destructive effect upon the illusion of tragic dignity our self-esteem had thrown over the contests of mankind with the sea. On that exquisite day of gently breathing peace and veiled sunshine perished my romantic love to what men's imagination had proclaimed the most august aspect of Nature. The cynical indifference of the sea to the merits of human suffering and courage,

laid bare in this ridiculous, panic-tainted
performance extorted from the dire extremity of
nine good and honourable seamen, revolted me.
I saw the duplicity of the sea's most tender
mood. It was so because it could not help itself,
but the awed respect of the early days was gone.
I felt ready to smile bitterly at its enchanting
charm and glare viciously at its furies. In a
moment, before we shoved off, I had looked
coolly at the life of my choice. Its illusions
were gone, but its fascination remained. I had
become a seaman at last.

— JOSEPH CONRAD,
The Mirror of the Sea, 1906

The wreck of the Mildred,
Scilly Islands, April 1912.
The ship foundered in thick
fog and was left to her fate
by captain and crew.

THE ENDURING SEA

THE CONCLUSION OF ZOOLOGIST RACHEL CARSON'S THIRD BOOK, *THE EDGE OF THE SEA*, IS A MEDITATION ON "THE UNIFYING TOUCH OF THE SEA," WHICH DISSOLVES TIME AND DISTANCE AND MATTER. WRITTEN SEVEN YEARS BEFORE *SILENT SPRING*, HER FAMOUS ENVIRONMENTAL TREATISE, "THE ENDURING SEA" INEVITABLY MAKES ONE WONDER WHAT CARSON WOULD HAVE THOUGHT OF OUR EFFORTS TO STABILIZE — "HARDEN" WITH BULKHEADS AND JETTYS AND BREAKWATERS — THE COASTS WHOSE EVER-CHANGING FORMS SHE CELEBRATES HERE.

Rocky outer coast, Big Sur, California.

Now I hear the sea sounds about me; the night high tide is rising, swirling with a confused rush of waters against the rocks below my study window. Fog has come into the bay from the open sea, and it lies over water and over the land's edge, seeping back into the spruces and stealing softly among the juniper and the bayberry. The restive waters, the cold wet breath of the fog, are of a world in which man is an uneasy trespasser; he punctuates the night with the complaining groan and grunt of a foghorn, sensing the power and menace of the sea.

Hearing the rising tide, I think how it is pressing also against other shores I know — rising on a southern beach where there is no fog, but a moon edging all the waves with silver and touching the wet sands with lambent sheen, and on a still more distant shore sending its streaming currents against the moonlit pinnacles, and the dark caves of the coral rock.

Then in my thoughts these shores, so different in their nature and in the inhabitants they support, are made one by the unifying touch of the sea. From the differences I sense in this particular instant of time that is mine are but the differences of a moment, determined by our place in the stream of time and in the long rhythms of the sea. Once this rocky coast beneath me was a plain of sand; then the sea rose and found a new shore line. And again in some shadowy future the surf will have ground these rocks to sand and will have returned the coast to its earliest state.

And so in my mind's eye these coastal forms merge and blend in a shifting, kaleidoscopic pattern in which there is no finality, no ultimate and fixed reality — earth becoming fluid as the sea itself.

On all these shores there are echoes of past and future: of the flow of time, obliterating yet containing all that has gone before; of the sea's eternal rhythms — the tides, the beat of surf, the pressing rivers of the currents — shaping, changing, dominating; of the stream of life, flowing as inexorably as any ocean current, from past to unknown future. For as the shore configuration changes in the flow of time, the pattern of life changes, never static, never quite the same from year to year. Whenever the sea builds a new coast, waves of living creatures surge against it, seeking a foothold, establishing their colonies. And so we come to perceive life as a force as tangible as any of the physical realities of the sea, a force strong and purposeful, as incapable of being crushed or diverted from its ends as the rising tide.

Contemplating the teeming life of the shore, we have an uneasy sense of the communication of some universal truth that lies just beyond our grasp. What is the message signaled by the hordes of diatoms, flashing their microscopic lights in the night sea? What truth is expressed by the legions of the barnacles, whitening the rocks with their habitations, each small creature within finding the necessities of its existence in

the sweep of the surf? And what is the meaning of so tiny a being as the transparent wisp of protoplasm that is a sea lace, existing for some reason inscrutable to us — a reason that demands its presence by the trillion amid the rocks and weeds of the shore? The meaning haunts and ever eludes us, and in its very pursuit we approach the ultimate mystery of Life itself.

— RACHEL CARSON, *The Edge of the Sea*, 1955

Rising storm, Big Sur, California.

THE STAR THROWER

ANTHROPOLOGIST LOREN EISELEY FINDS HIMSELF ON A TROPICAL BEACH, WHERE "DEATH WALKS HUGELY AND IN MANY FORMS" AND ONE MADMAN IS A RESCUER. AS A SCIENTIST, HE RECOGNIZES THE UTTER FUTILITY OF THE STAR THROWER'S MISSION, BUT NONETHELESS FINDS IN HIS ACTIONS A MODEL FOR HUMAN BEHAVIOR.

The beaches of Costabel are littered with the debris of life. Shells are cast up in windrows; a hermit crab, fumbling for a new home in the depths, is tossed naked ashore, where the waiting gulls cut him to pieces. Along the strip of wet sand that marks the ebbing and flowing of the tide death walks hugely and in many forms. Even the torn fragments of green sponge yield bits of scrambling life striving to return to the great mother that has nourished and protected them.

In the end the sea rejects its offspring. They cannot fight their way home through the surf which casts them repeatedly back upon the shore. The tiny breathing pores of starfish are stuffed with sand. The rising sun shrivels the mucilaginous bodies of the unprotected. The sea beach and its endless war are soundless. Nothing screams but the gulls.

In the night, particularly in the tourist season, or during great storms, one can observe another vulturine activity. One can see, in the hour before dawn on the ebb tide, electric torches bobbing like fireflies along the beach. It is the sign of the professional shellers seeking to outrun and anticipate their less aggressive neighbors. A kind of greedy madness sweeps over the competing collectors. After a storm one can see them hurrying along with bundles of gathered starfish, or, toppling and overburdened, clutching bags of living shells whose hidden occupants will be slowly cooked and dissolved in the outdoor kettles provided by the resort hotels for the cleaning of

specimens. Following one such episode I met the star thrower.

As soon as the ebb was flowing, as soon as I could make out in my sleeplessness the flashlights on the beach, I arose and dressed in the dark. As I came down the steps to the shore I could hear the deeper rumble of the surf. A gaping hole filled with churning sand had cut sharply into the breakwater. Flying sand as light as powder coated every exposed object like snow. I made my way around the altered edges of the cove and proceeded on my morning walk up the shore. Now and then a stooping figure moved in the gloom or a rain squall swept past me with light pattering steps. There was a faint sense of coming light somewhere behind me in the east.

Soon I began to make out objects, upended timbers, conch shells, sea wrack wrenched from the far-out kelp forests. A pink-clawed crab encased in a green cup of sponge lay sprawling where the waves had tossed him. Long-limbed starfish were strewn everywhere, as though the night sky had showered down. I paused once briefly. A small octopus, its beautiful dark-lensed eyes bleared with sand, gazed up at me from a ragged bundle of tentacles. I hesitated, and touched it briefly with my foot. It was dead. I paced on once more before the spreading whitecaps of the surf.

The shore grew steeper, the sound of the sea heavier and more menacing, as I rounded a bluff into the full blast of the offshore wind. I was away

Sea star in a sea-grass bed off Belize, in the Caribbean. The red mangroves in the distance grow throughout the tropics, in mud flats.

from the shellers now and strode more rapidly over the wet sand that effaced my footprints. Around the next point there might be a refuge from the wind. The sun behind me was pressing upward at the horizon's rim — an ominous red glare amidst the tumbling blackness of the clouds. Ahead of me, over the projecting point, a gigantic rainbow of incredible perfection had sprung shimmering into existence. Somewhere toward its foot I discerned a human figure standing, as it seemed to me, within the rainbow, though unconscious of his position. He was gazing fixedly at something in the sand.

Eventually he stooped and flung the object beyond the breaking surf. I labored toward him over a half mile of uncertain footing. By the time I reached him the rainbow had receded ahead of us, but something of its color still ran hastily in many changing lights across his features. He was starting to kneel again.

In a pool of sand and silt a starfish had thrust its arms up stiffly and was holding its body away from the stifling mud.

"It's still alive," I ventured.

"Yes," he said, and with a quick yet gentle movement he picked up the star and spun it over my head and far out into the sea. It sank in a burst of spume, and the waters roared once more.

"It may live," he said, "if the offshore pull is strong enough." He spoke gently, and across his bronzed worn face the light still came and went in subtly altering colors.

"There are not many come this far," I said, groping in a sudden embarrassment for words. "Do you collect?"

"Only like this," he said softly, gesturing amidst the wreckage of the shore. "And only for the living." He stooped again, oblivious of my curiosity, and skipped another star neatly across the water.

"The stars," he said, "throw well. One can help them."

He looked full at me with a faint question kindling in his eyes, which seemed to take on the far depths of the sea.

"I do not collect," I said uncomfortably, the wind beating at my garments. "Neither the living nor the dead. I gave it up a long time ago. Death is the only successful collector." I could feel the full night blackness in my skull and the terrible eye resuming its indifferent journey. I nodded and walked away, leaving him there upon the dune with that great rainbow ranging up the sky behind him. . . .

Man is himself, like the universe he inhabits, like the demoniacal stirrings of the ooze from which he sprang, a tale of desolations. He walks in his mind from birth to death the long resounding shores of endless disillusionment. Finally, the commitment to life departs or turns to bitterness. But out of such desolation emerges the awesome freedom to choose — to choose beyond the narrowly circumscribed circle that delimits the animal being. In that widening ring of human choice, chaos and order renew their symbolic struggle in the role of titans. They contend for the destiny of a world.

Somewhere far up the coast wandered the star thrower beneath his rainbow. Our exchange had been brief because upon that coast I had learned that men who ventured out at dawn resented others in the greediness of their compulsive collecting. I had also been abrupt because I had, in the terms of my profession and experience, nothing to say. The star thrower was mad, and his particular acts were a folly with which I had not chosen to associate myself. I was an observer and a scientist. Nevertheless, I had seen the rainbow attempting to attach itself to earth.

On a point of land, as though projecting into a domain beyond us, I found the star thrower. In the sweet rain-swept morning, that great many-hued rainbow still lurked and wavered tentatively

28

beyond him. Silently I sought and picked up a still-living star, spinning it far out into the waves. I spoke once briefly. "I understand," I said. "Call me another thrower." Only then I allowed myself to think, He is not alone any longer. After us there will be others.

We were part of the rainbow — an unexplained projection into the natural. As I went down the beach I could feel the drawing of a circle in men's minds, like that lowering, shifting realm of color in which the thrower labored. It was a visible model of something toward which man's mind had striven, the circle of perfection.

I picked and flung another star. Perhaps far outward on the rim of space a genuine star was similarly seized and flung. I could feel the movement in my body. It was like a sowing — the sowing of life on an infinitely gigantic scale.

I looked back across my shoulder. Small and dark against the receding rainbow, the star thrower stooped and flung once more. I never looked again. The task we had assumed was too immense for gazing. I flung and flung again while all about us roared the insatiable waters of death.

But we, pale and alone and small in that immensity, hurled back the living stars. Somewhere far off, across bottomless abysses, I felt as though another world was flung more joyfully. I could have thrown in a frenzy of joy, but I set my shoulders and cast, as the thrower in the rainbow cast, slowly, deliberately, and well. The task was not to be assumed lightly, for it was men as well as starfish that we sought to save.

— LOREN EISELEY,
The Unexpected Universe, 1969

BEACH WALK

THE WAVES AND WILDLIFE OF AN OCEAN BEACH ARE NOT MERELY PARTS OF A PICTURESQUE LANDSCAPE; THEY ARE DYNAMIC EXPRESSIONS OF THE FORCES OF NATURE. A PHYSICIST AND A NATURALIST WALK TOGETHER ON A BEACH AND EACH ONE USES THE LENSES OF SCIENCE TO PENETRATE BEHIND FAMILIAR APPEARANCES AND BEGIN TO GRASP ESSENTIAL PROCESSES.

Let it be clear that I am no slouch when it comes to beaches. I have walked them summer and winter, in sunshine and in horizontal rain that felt like bird shot, on days when the sea was glassy and days when a storm threatened the land. I have been there body surfing, birding, surf casting, gathering seaweed for mulch, beachcombing for treasure, or just walking. I feel as at home on a beach as any "civilized" person can feel in a wild place.

Yet a single six-mile stroll on a warm day last spring opened my eyes to a whole new way of seeing the beach, of understanding and even quantifying what is happening there. Walking with me was Jim Trefil, a physicist at the University of Virginia who has a knack of seeing the extraordinary in the commonplace. . . . That day, each of us was going to teach the other something about the way he thought about what he saw.

The sea is calm and waves breaking on the beach are a foot to a foot-and-a-half high. Even that gentle a surf, however, stirs up the sand. Jim points out to me that the waves are not arriving perpendicular to the beach, but rather at a slight angle. The water comes at us from a bit to the left and washes back a bit to the right. He starts counting the time between waves and finds that one breaks about every ten seconds. He stands in the wash, watching individual grains of sand move, doing calculations in his head. If a grain of sand is moved only an eighth of an inch by each wave, he reasons, then about 9,000 waves a day could move it more than 90 feet, amounting to several miles a year.

We walk farther. I keep using binoculars to look up the beach, counting willets, grackles, ring-billed gulls. Jim is pointing out that incoming waves break when their height is slightly more than three-fourths the depth of the water. They break in different ways, depending on the configuration of the ocean floor and on how much faster their tops are going than their bottoms. We move up into the dry sand near the dunes, where I spend my time trying to follow raccoon tracks until he shows me the color patterns in the wind ripples: because of their varying masses, grains of different minerals are transported differently by the moving air.

Farther up we stop for lunch. Jim is thinking about the falling tide and how the tidal bulge in the ocean caused by the moon cannot keep up with the moon as we spin underneath. He believes you can see the basic forces that run the Universe anywhere, even the beach. I watch terns plunging into the water just beyond the surf and imagine I can feel the pull of the moon on the Earth beneath me.

All is not gravitation and energy. Sitting there I suddenly realize that some of the "willets" probing the sand have long bills that curve downward. Their heads are striped fore-to-aft. They are whimbrels, refueling on their way to Arctic breeding grounds. So far Jim has paid no attention to the birds along the beach, but he is interested

Clammers on the beach, Moclips, Washington, 1982.

31

in the whimbrels and how far they migrate. He becomes fascinated with birders, people who will travel hundreds of miles to see one bird, and decides he may begin a life list. It is he who first spots two American oyster-catchers — flashes of bold black and white with heavy, long red bills — flying north along the surf line. Later, I log his sighting at the ranger's station to give his list an official start.

Back down the beach. I am still looking at birds, Jim concentrating on the waves. When conditions are just right, he tells me, a hydraulic jump occurs. This can happen when a large wave has finished its run up the beach and the water is flowing back toward the surf. The next wave breaks, and the water in that wave is headed for shore, but it is moving on top of water that's moving rapidly the other way. Finally the new wave can make no further forward progress, but boils and churns in place, a stationary wave on top of the receding wash.

In the meantime the wind has shifted around so it is blowing out of the south at about 15 knots. It is blowing straight up the beach, parallel to the surf line. Jim points out to me that within half an hour a whole new pattern of ripples has appeared in the dry sand. I had never paid much attention to the ripples. To me they were just the unseen background against which I looked for living things.

Farther down I do see a living thing, a duck-size black bird plumped down on the wet sand just above the reach of the waves. As we come closer I see the white patches on the front and back of the head and the lumpy orange-and-red bill. The surf scoter stares at us with what seems a distinctly jaundiced eye until we get to within 15 feet, then flies heavily out over the water, red legs dangling, to land just beyond the surf. I am excited at having been so near a bird I had never seen up close before; Jim, who had held back to give me the best chance at getting close, asks why I wanted to do so. Up to now, the birds have been just part of the unseen background for him, he says, transients across the stage where energy and gravity, fluid mechanics and surface tension play out their parts.

He has plenty to look at: the foam left by a receding wave is a sheet of small bubbles, iridescent from the right angle, sparkling when seen against the sun. Sometimes I watch them the way I watch a fire dreamily, out of focus. Jim watches them shrink, water running down their sides, until the air pressure inside equals the surface tension shrinking them and the bubble reaches equilibrium. Water droplets, which act much like bubbles, became the model for Niels Bohr's description of the atomic nucleus, one of the first coherent explanations of the atom's core, Jim tells me. The strong force binding the neutrons and protons together is the analogue of the surface tension that condenses a bubble or water droplet, while the electric force, which repels like-charged particles inside the nucleus, is the analogue of the pressure inside such spheres that resists further shrinking.

Bursting bubbles inject matter into the atmosphere, plankton and algae as well as salt. As the air pressure inside increases, the bubble pushes down into the water. When the bubble breaks, the pressure is released and the depressed portion of the water surface springs back toward normal. It actually goes too far and a column rises up where the depression was. The rising column breaks into drops some of which are ejected as much as a foot into the air. The drops then evaporate, leaving the plankton and algae to be carried by the wind. The quantities involved are enormous: foam and its bubbles form not just where waves are breaking alongshore, but anywhere waves are breaking in the open sea (whitecaps). At any one moment, an area of the Earth's surface roughly the size of North America is covered with foam. In a year,

LEFT GUTTING SALMON
ON THE PROCESSING
LINE, SAKHALIN ISLAND,
RUSSIA, 1990. THREE
SHIFTS WORK AROUND
THE CLOCK DURING
THE PEAK SALMON RUN.
PROCESSORS ARE PAID
ACCORDING TO THEIR
TEAM'S PRODUCTION.

ABOVE EXHAUSTED
WORKERS TAKE A BREAK.
SOME SHIFTS ARE
TWELVE-TO-FOURTEEN
HOURS LONG.

MA'KULIWA

To go to sea is to surrender to the whims of nature, to place one's life at the mercy of forces always beyond one's control and often beyond one's imagination. But because it is human to try to control one's own destiny, seafarers the world over practice rituals intended to influence nature, to encourage it to be benign rather than malevolent. There are prayers, talismans, amulets, offerings, and, in some societies, elaborate ceremonies designed to placate the sea and encourage the fish. "Whatever you do," an Indonesian flying fisherman says, "flatter them, feed them sweet things, talk respectfully to them." But a switch from subsistence to export markets tends to drive out old customs: the rituals described here are waning, as Mandar fishermen have begun to fish with gill nets for the lucrative Japanese market, which imports flying-fish roe for sushi.

A Mandar fishing-boat captain faces Mecca and prays. The boat will be tied to a raft that is designed to attract schools of fish, which the crew will catch using a purse seine net.

During the month of April on the Mandar coast of Indonesia, a two-hundred-mile-long stretch of poor fishing villages squinched between South Sulawesi's mountainous interior and the 5,000- foot depths of the Makassar Strait, Mandar fishermen draw their narrow, double-outrigger boats onto shore, to prepare them for the flying fish season. "The flying fish," my Mandar friend, Pak Nuhung, often reminded me, "is an extraordinary fish! No other fish in the world is in need of such flattery, cajoling, enticement. This fish must be seduced into entering our *buaro* (traps)." To capture this extraordinary fish, and to protect the boat from human as well as natural hazards, many Mandar fishing crews seek to render their vessels invulnerable and beautiful through elaborate ritual operations: tying and retying all critical structures on the boat, particularly the outriggers and the rudder, stashing small clusters of especially powerful plants known as "the boat's medicine" within critical joints, and through performance of *ma'kuliwa tui-tuing*. The latter is an embarkation ritual which simultaneously repels magical assaults and increases the potency and attractiveness of the vessel to invisible, watchful spirits believed to control the movements of the flying fish and all other creatures which swim the seas. During ma'kuliwa, the vessel is likened to a "young woman wearing make-up," an animate, alluring presence on the seas. The account of ma'kuliwa which follows is based on my field notes in Luaor village, on the day before the *Kurniawan*, Pak

Nuhung's father's fishing boat, embarked in 1989.

Not all dangers come from the sea. Threats to the boat and its crew also arise from within the hearts of men. They are set in motion through magical practices including attacks launched "like arrows," in the sands below the boat's keel. If these attempts succeed, they will penetrate and permeate the boat, affecting the crew and its fate at sea. Performance of *ma'sulo,* or "torching," during ma'kuliwa, aims at repelling human magical assaults. Completion of *ma'cobo,* or "palming," is directed at attracting and beguiling those spirits who direct the movements of *tui-tuing,* the flying fish.

It is pitch dark. The time is between seven and eight-thirty at night. The action is on the shore, around and near the boat. In complete silence, on the beach behind the *Kurniawan,* Abdul Raup, the lanky, laconic ritual practitioner, or *sando pa'tui-tuing,* lights a torch of rattan vines about seven feet long and walks toward the boat anchored in the water. Beginning at the stern, the sando touches, grazes, and traces the torch around the side of the boat, at a point just above where sea water laps against the hull.

Raup circumambulates the boat, ringing it and touching it with fire, rendering it invulnerable as he walks. From the right-hand side of the stern, he wades out into the water of the lagoon, touching the hull with the torch and moving toward the sea-ward pointing bow. I wait for him on the other side of the boat in the darkness. I can hear

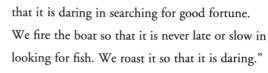

the popping sounds of his torch and see puffs of grey smoke rising from somewhere beyond the bow, before I see him. And then he comes around the side where I am waiting, the fire-shaman, with his huge, deep-set eyes and ambling gait. Raup carries his sizzling torch of vines, steps nimbly around the prow, bends with his torch below the outrigger arm, and passes me, leaving a line of ash along the *Kurniawan*'s hull, a light robin's egg blue.

At dawn the following day, these grey and salt-and-pepper-flecked splotches of ash, laid down at night, will be inspected for the portents they carry concerning the fate of the crew. Abdul Raup, master of the adze and drill, boat builder, and sando, completes three circumambulations of the *Kurniawan*, extinguishes the torch and walks a few yards down to Jahmal's boat, a neighbor's craft which will embark tomorrow with the *Kurniawan*. The sky is utterly black. The water is calm and as dark as the heavens. As Raup begins his circling and firing again, I follow his movements by the reflections of his torch in the darkened waters. I ask Nuhung's father, Ne'Tata, the *Kurniawan*'s captain, "Why is the boat being torched?" and Ne'Tata answers unceremoniously: "Because we want to catch a lot of fish. We torch the boat so

that it is daring in searching for good fortune. We fire the boat so that it is never late or slow in looking for fish. We roast it so that it is daring."

Up on deck the following evening, some cosmic cooking is about to take place. Raup is beginning to prepare an invulnerability porridge for a procedure known as *ma'cobo,* or "palm printing." Assisted by Ne'Tata, Raup mixes up a potent brew of powerful plants — *dui-dui, lere-lere'* and *bea'-bea'.* The names, forms, and fragrances of these plants resonate with and embody the desired outcomes of flying fishing: ideas of many, of sticking in the net, of many things being caught. Climbing up on board and seating himself on the right side of the deck, Raup places a large bowl of rice flour and small piles of fragrant plants before him. He reaches over to Ne'Tata, takes a chicken egg and breaks it over the flour, and begins to knead the mixture with both hands. Every now and then he calls for some extra something — a little *minyak lommo',* the sweet, fragrant red oil, or a little something else, which he silently adds to the potent, yeasty-looking batch. A glass containing water from the mosque's well is poured into the mixture. The small light of a wick lamp illuminates these unearthly cooking operations.

ABOVE *Uvo, Solomon Islands, 1970s. This shark priest's fishing float was used in a ritual to catch flying fish to propitiate ancestor spirits. Such floats were baited with hermit crabs and set out about a dozen at a time.*

RIGHT *Spirit protectors, Nunivak Island, Alaska, 1927. These wooden plaques of a frowning woman with chin tattoos and a smiling man, perhaps depicting spirits, would have been placed inside kayak cockpits by Inuit hunters to serve as protective charms.*

For good fishing, Pete Blackwell kisses the first sockeye salmon of the season before throwing it overboard, Bristol Bay, Alaska, 1991.

Abdul Raup inspects the batch and finds it ready. He moves toward the forward hatch, and, holding a small oil-wick lamp in his left hand, steps down into the darkness of the hold. From the deck, Ne'Tata hands Raup the *co'bo* mixture and a small brazier of incense. Looking down into the hold, I can make out the sando's movements, despite the almost complete darkness. He wafts the musky smoke of the censor toward the co'bo mixture, incensing it. Then he waves the fragrant smoke toward the boards on the sides of the boat. Incense is also propelled toward the boat's "navel," or central mast, called the *posi tiang,* where a guardian spirit is believed to dwell.

Raup dips his right index finger into the co'bo mixture and, with a small coating of the magic plant paste on his finger tip, in complete and utter darkness, he touches, from right to left, his temples, his forehead, his eyes, and his nose, leaving a deposit of co'bo paste on each bodily center. He unbuttons his shirt and daubs co'bo paste on his belly button and then looks up (directly into my face as I am crouching above the open hatchway, watching his every move) and, opening his mouth so widely that he looks as if he is grimacing, he touches the upper rear of his throat, "the place that never runs dry," with co'bo paste. Once his body has been incensed and daubed with co'bo paste, Raup touches the main post and the interior sides of the hull with the mixture, beginning the process of making the boat invulnerable from within the interior of the craft and moving up and outward.

Once outside and wading in the water, his bowl of co'bo mixture in his left hand, clutched against his chest, and moving from the stern to the bow, Raup walks alongside the *Kurniawan,* daubing its sides, outrigger arms, and outriggers with co'bo paste. His movements are deliberate and methodical, there is nothing hurried about this operation. Every foot or so, he stops, places his right hand into the sticky plant porridge, reaches out toward the hull where he places his

palm and outstretched fingers. As he lifts his hand from the hull, an imprint of his palming, a sign of invulnerability, as well as intense, radiant beauty, is impressed upon the hull. Like the ash from the torching firewalk, Raup's second ritual operation creates a ring of patches and prints on the boat's body, a magic paste of potent plants and palm prints which will also be scrutinized intensely the following morning, read for signs of this ship's future, or *dalle,* at sea.

As he emerges from the hold of the *Kurniawan,* Raup turns toward me on the deck and says, "I'll co'bo you! Want it?" "Sure," I reply. He dips his hand into the co'bo mixture and daubs three points, one on my right temple, one on the center of my forehead, and one on my left temple. On deck, he stands back from his handiwork, and in the dark, alone on the boat with me, he says with a smirk, "Your woman will like it!" Daubed with the signs of invulnerability and beauty, torched and made up as young women, the *Kurniawan* and I are ready to sail for flying fish.

— CHARLES ZERNER, 1994

CHARMS AND RITUALS

RISKING DEATH DAILY, FISHERMEN AROUND THE WORLD TURN TO CHARMS, AMULETS, OR TOTEMS AS PROTECTIVE DEVICES. IN SOME CASES, THESE OBJECTS ARE WORN BY AN INDIVIDUAL; IN OTHERS, THEY SERVE AS SHRINES FOR A CREW, FAMILY, OR COMMUNITY AS A WHOLE.

WEATHER CHARM, CAROLINE ISLANDS, MICRONESIA. NAVIGATORS IN CANOES TRAVELING AMONG THE MICRONESIAN ISLANDS USED THESE CHARMS TO DIVERT STORMS AND BAD WEATHER, AND TO APPEAL TO BENEVOLENT WATER SPIRITS. WEATHER CHARMS WERE ALSO USED IN SORCERY.

RIGHT STATUETTE OF OUR LADY OF GOOD VOYAGE, NATIONAL SHRINE OF THE FISHERMAN, GLOUCESTER, MASSACHUSETTS. THE MOST COMMON CHARMS AMONG NORTH ATLANTIC FISHERMEN ARE RELIGIOUS MEDALS, PORTUGUESE FIST-OF-POWER FIGURES, OR FAMILY MEMENTOS PASSED FROM FATHER TO SON.

OPPOSITE TEMPLE FOR THE SAFE RETURN OF FISHERMEN, BALI.

EMPHASIZE THE UNITY OF
FISHING COMMUNITIES,
COMMEMORATING LOSSES
AND CELEBRATING SUC-
CESSES.

OPPOSITE A VEZO
TURTLE HUNTER ADORNS
HIS CANOE BOW WITH
TURTLE BLOOD TO HONOR
THE PREY, MADAGASCAR.
THE VESO, MOSTLY
SUBSISTENCE FISHERS,
CALL THEMSELVES
"THE PEOPLE WHO KNOW
THE SEA."

NAUTICAL SAYINGS

The sea has contributed countless words, expressions, and concepts to mainstream English. We seldom reflect upon their oceangoing origin when we use such expressions as all in the same boat, blow over, crabby, don't give up the ship, down the hatch, getting the drift, happy as a clam, keel over, like a fish out of water, making headway, navy blue, small fry, the coast is clear, go off the deep end, take the wind out of his sails, or wide berth. Other expressions with less obvious nautical roots include:

GROGGY comes from "grog," the name sailors in the British Royal Navy disdainfully used for their daily ration of a half pint of rum, after it was decreed in 1740 that the rum should be diluted with an equal amount of water. The unpopular order was issued by Vice Admiral Sir Edward Vernon, nicknamed "Old Grog" because of the impressive grogram cloak he wore on deck.

The expression "to know the ropes" originally referred to the rigging of a ship. A three-masted sailing ship may have as many as one hundred and fifty separately named lines, from the royal stay (which runs from the top of the foremast to the tip of the bowsprit) to the spanker boom lift (which holds up the spanker boom). Only the savviest sailor knows them all.

HORSE LATITUDES refer to the regions of calm found at latitudes 30°N. to 30°S. It is said that sailing ships carrying horses to America, when becalmed in these latitudes, had to throw them overboard in order to lighten their vessels and take advantage of any gentle breezes that might blow their way.

OVERWHELM comes from the Middle English word meaning "to capsize."

POOPED OUT originally described the condition of seamen caught on the poop or aft deck when a wave from a following sea crashed down upon it.

RUMMAGE SALE stems from the French word *arrimage,* meaning "the loading of a cargo ship." Damaged cargo was occasionally sold at special warehouse sales.

SKYSCRAPER traditionally referred to the topsail of a ship, and only more recently has come to mean a tall building.

SLUSH FUNDS were once the personal funds of ship cooks, who earned them by skimming off the fat, or "slush," from cooking and selling it when the ship came into port.

STRANDED vessels were ones that had drifted or run aground on a strand, or beach.

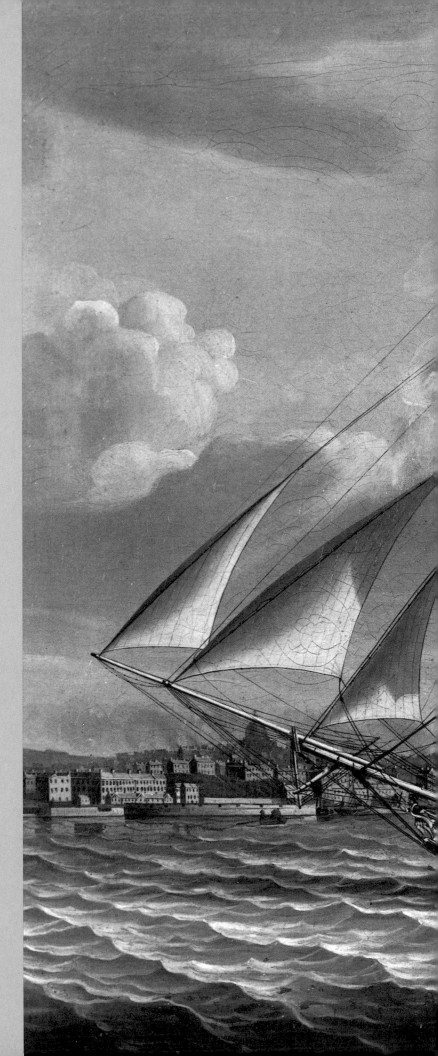

A HARD LIFE

EVERY PROFESSION EXACTS A PRICE FOR ITS SATISFACTIONS. SOME REQUIRE THE SACRIFICE OF INDEPENDENCE, SOME OF FAMILY TIME OR CREATURE COMFORTS OR SAFETY OR SECURITY. BUT FEW PROFESSIONS ARE AS DEMANDING AS FISHING. THE HOURS ARE LONG AND IRREGULAR. WORKING CONDITIONS RANGE FROM DIFFICULT TO APPALLING AND SUCCESS IS UNCERTAIN. AND YET FISHING HAS ITS REWARDS FOR THOSE WHO INSIST ON PURSUING IT. WRITER PETER MATTHEISSEN BEFRIENDED THE TIGHT-KNIT FAMILIES WHO, GENERATION AFTER GENERATION, SOME FOR NEARLY TWO HUNDRED YEARS, HAVE BEEN FISHING THE WATERS OFF THE SOUTH FORK OF LONG ISLAND. WHY DO THEY PERSIST? REDUCED TO ONE WORD, THE ANSWER IS INDEPENDENCE. BUT IT IS AN INDEPENDENCE TEMPERED BY A STRONG SENSE OF COMMUNITY AND PRIDE IN SHARED PURPOSE AND ACCOMPLISHMENT.

South Fork fishermen, Long Island, 1982.

That spring of 1955, I knew my job and I enjoyed it. Lying back on the damp nets stacked in the dory after the early morning sets, trundling along the bright white sand in the soft spring light, enjoying the spring breath of the fertile loam that came all the way down to the back of the dunes and blew out over the clean beach — the springtime filled me with well-being despite my drowsiness, despite sore hands, despite the prospect of a long hard day that, if we were lucky, would end long after dark in the cold freezer of Montauk Seafood, hosing down and packing and icing bass.

As a seasonal fisherman, I could afford to romanticize this life, to indulge myself in wondering why a veteran bayman such as Milt Miller would inveigh so vehemently against his lifelong occupation. (Not that this habit was confined to Milt. As an old saying goes, "I'm gonna put this goddamn oar over my shoulder and head west, and the first sonofabitch asks me what it is, that's where I stick it in the ground and settle.") Like most baymen, Milt was a skilled carpenter, boat-builder, and jack-of-all-trades who could easily have set up his own shop or found well-paid work in the construction industry; no law but his own had told him to be a fisherman. But as he said, it was "in his blood," there was nothing to be done about it.

In spring the net is set from east to west, insuring a hard row into the onshore wind each afternoon, and Milt, behind me at the bow oars,

would curse into my ear with every heave. "My boy Mickey ever touches a fishin net, I'll tan his hide! Worked like a donkey all my life and here I am, still workin like a donkey, cause I don't *know* nothin! Not one bit better off than when I started!" Milt would row some, get his breath, and start all over. "Boy, this life here is just one big mess, in case you ain't found that out yet! Best thing Mickey could do would be read up a little, get a job as a psycho-analyst, you know, like that loony I run into over here the other day. What you do, you lay down there on a couch and earn you a day's pay in a hour, just listenin to rich people belly-achin! You and Johnny got a education, Pete, how come *you* ain't psycho-analyzin? Pay you better than bassin, I'll tell you that!" Milt would be winking slyly at Stewart Lester in the stern, I could read it in the grin on Stewart's face, but he was serious, too: he was fighting to make sense of a hard life.

Milt said that in the early thirties, when he had fished Captain Gabe Edwards's set nets on a share basis, both bass and bluefish had been very scarce. "Cap'n Gabe used to talk about acres of bluefish back in the old days, and who was I to disbelieve a captain and deacon of the church who never was known to tell anything but the truth? But if we got four or five bluefish out of eight or ten set nets, we thought we were lucky." He stuck out his left thumb to show the scar. "That was done by a bluefish when I was just a kid. He was slippin out of the net, and because he was so

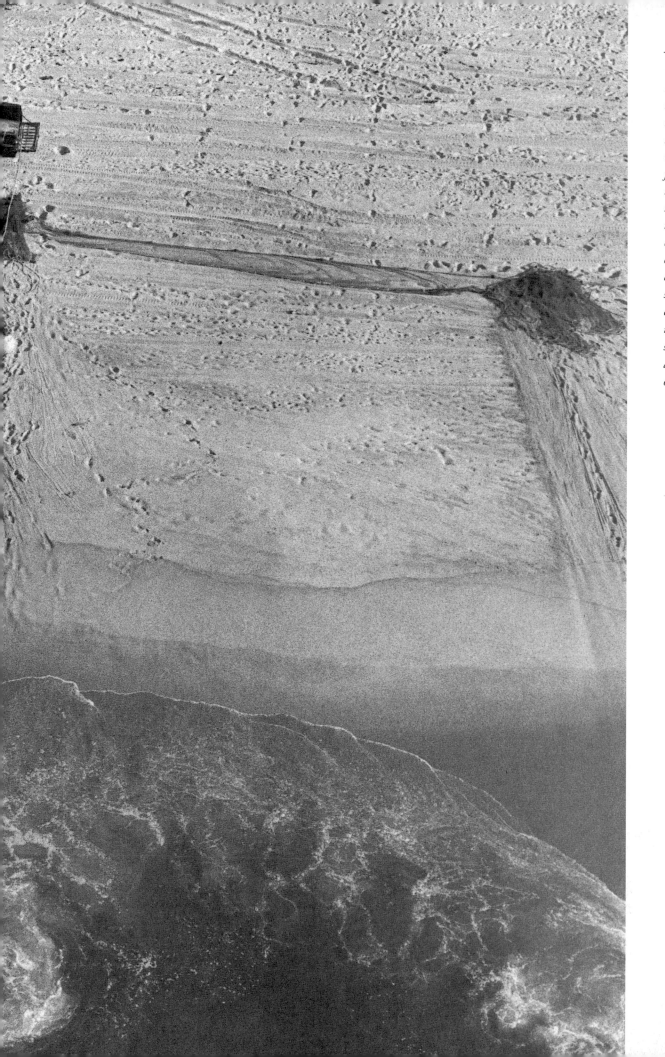

Ocean haul seining is among the most physically demanding forms of fishing. The net is set by rowing a boat into the crashing surf and winched back onto the beach from trucks. It is not until the net is almost landed that the crew can see whether there is a good catch in the bag. This 1983 aerial view of Pete Cromer's haul seine crew on a Long Island beach shows the moment at the end of the run when the net is being hauled up on the sand and the dory has already been winched up on its trailer.

rare y'know, I wasn't goin to let him go; stuck my hand down quick, right into his mouth, and when I flung him, them teeth opened me up right to the bone."

In 1934 — he remembers the year because of the intense cold that winter, the last winter in memory when the salt water froze solid on Gardiners Bay, permitting walking expeditions from Fireplace to Gardiners Island — Milt had hauled seine with his cousin Elisha Ammon "down around Mecox," where his father was caretaker of the Mecox lifeboat station. Their rig was a Model T Ford with oversized tires and a platform for a small boat, and the small net was hauled by a winch built by Elisha from parts of an old potato digger, turned by the drive shaft at the front end of the car. So far as Milt knows, this was the first winch ever used for hauling on the beach. "And Model T's were the first real beach vehicles," he remembers. "Before that they'd row down the beach to where they spotted fish, set net and haul, and bury the catch up in the sand, pickin 'em up later with the dory; then they'd row 'em back and hand-barrow 'em up to the roads at the beach landin."

Elisha Ammon, born in Springs in 1911, had spent most of his life out on Montauk, where he can recall killing five deer in a day for winter meat. His grandfather, born at sea and landing at Sag Harbor when what was then known as "the Port" was a port of entry, became a crewman on "the bony boats" — the early steamers fishing for the bony fish, menhaden. Later he worked on the traps and draggers for the Jake Wells and Parsons fish companies on Fort Pond Bay. Elisha's father had done the same, and Elisha did it, too; as he says himself, "I was never brang up to do anything else." With his Uncle Royce, he ran a string of 750 lobster pots, using dabs (daylight flounder) for bait, and for two years he worked in the Parsons garage to save up money before he got "brave enough to buy a boat"; it was here that he assembled the prototype winch used for hauling seine with the Model T at Mecox.

Elisha became a successful draggerman and swordfisherman, which meant that he was mostly away from home. "You'd just say 'Good-bye' when you leave and 'Here I am' when you get back. I'd never say when I was getting back; better just to say, You'll see me when I get home. Get up three o'clock, get home ten, eleven; children never even knew who I was until they was twelve-fourteen years old." Like many fishermen's sons in recent decades, Elisha's son quit fishing to become a carpenter when he got married, and although Elisha understands this, he is wistful. "Might still have a boat today," he says, "if he just stayed with me."

When he was fishing with Elisha, Milt had already married Etta Midgett, and in 1934 he moved to her home state of North Carolina. But his wife missed East Hampton, where her uncle and father had already established themselves as fishermen, and after three or four months the Millers returned to the South Fork. It was now the Depression. . . . "Times was so hard," Milt sighs today, "that what I done was row at night from Bonac Creek across to Gardiners and around to the east shore and sneak across the beach into Great Pond when there was no moon and scratch up maybe four bushel of clams, then lug them four bushel back over the beach out to my sharpie. And I couldn't afford them nice wire baskets, I used wood produce baskets with sharp wire handles, nearly cut my hands off. Then I'd row all the way back, and a couple of times I damn near swamped, comin around out of the lee at Cartwright Shoal. And when I was done, I got fifty cents a bushel for them clams, and was glad to have it; I had kids cryin at home, and I had no choice.

"Independence costs you a lot of money," he told me quietly. "I starved myself to death for independence when I could have made good money at a trade. You ever seen anybody yet get fired from fishin? No, no! You're just glad to find somebody stupid enough to go fishin *with* you."

— PETER MATTHIESSEN, *Men's Lives,* 1986

65

COMMUNITY

FISHING COMMUNITIES ARE OFTEN BOUND
BY TRADITIONS AND A STRONG SENSE
OF UNITY. CUSTOMS IN MANY TRADITIONAL
CULTURES REINFORCE COOPERATION IN
DIVIDING UP FISHING AREAS, THE CATCH,
AND THE WORK LOAD.

MAINE LOBSTERMEN
DIVIDE FISHING AREAS
THROUGH INFORMAL
AGREEMENTS, AND MARK
THEIR TERRITORIES BY
TYING EACH TRAP TO A
BUOY PAINTED WITH
AN INDIVIDUAL'S COLORS.
LONG-TIME RESIDENTS
CLAIM AND DEFEND THEIR
TERRITORIES FROM NEW-
COMERS AND FISHERMEN
FROM OTHER HARBORS.
FAMILY RELATIONSHIPS
ARE EXTREMELY IMPOR-
TANT IN DECIDING WHO
IS ALLOWED TO FISH AND
WHERE LOBSTERING
TAKES PLACE.

RIGHT LOBSTERMEN
MEETING IN THE FISH
HOUSE, MONHEGAN
ISLAND, MAINE.

PASSING ON KNOWLEDGE

FISHERMEN LEARN ABOUT WHERE TO FISH
AND HOW TO GET THERE THROUGH EXPERI-
ENCE AND BY WATCHING OTHER FISHERMEN.
MUCH OF THIS KNOWLEDGE IS PASSED
FROM FATHERS TO SONS.

ABOVE BEFORE DAWN,
A FATHER TEACHES
HIS SEVEN-YEAR-OLD SON
HOW TO SET LOBSTER
TRAPS, MONHEGAN
ISLAND, MAINE.

RIGHT A FATHER
SHOWS HIS SON HOW TO
BUILD AN OUTRIGGER
CANOE, MADAGASCAR.

A ROTTEN SHIP

NINETY-FIVE PERCENT OF THE
WORLD'S FREIGHT TRAVELS ON
THE SEA, IN SHIPS AND WITH
CREWS FROM NEARLY EVERY
NATION. AND YET LESS THAN
FIVE PERCENT OF AMERICAN
CARGO IS CARRIED IN AMERICAN
SHIPS. THE UNITED STATES
MERCHANT MARINE HAS BEEN
IN DECLINE FOR MANY YEARS,
BATTERED BY COMPETITION
FROM SHIPS THAT CAN PLY THE
OCEANS MORE CHEAPLY, UN-
BURDENED BY COSTLY SAFETY
REGULATIONS AND HIGHLY
PAID SEAMEN. STILL, THE SEA
LURES THE AMERICAN SAILOR.
AS SHIPS GET BIGGER AND
CREWS SMALLER, HE COMPETES
FIERCELY FOR THE FEW AVAIL-
ABLE JOBS, WAITING RESTLESS-
LY ON SHORE FOR MONTHS UNTIL
THE DATE ON HIS UNION CARD
ASSURES HIM A BERTH ON THE
NEXT BOAT OUT, WHICH COULD
BE GOING ANYWHERE, FROM
A NEARBY PORT TO THE OTHER
SIDE OF THE WORLD. WRITER
JOHN MCPHEE MET PETER FRITZ
ON A FORTY-TWO DAY VOYAGE
ABROAD THE MERCHANT SHIP
S.S. *STELLA LYKES*.

OPPOSITE *Without
a survival suit, a person
overboard will succumb to
hypothermia in minutes in
near-freezing Alaskan waters.*

RIGHT
Iced-over crab boat, Alaska.

On February 11, 1983, a collier called Marine Electric went out of the Chesapeake Bay in a winter storm with a million dollars' worth of coal. She was a ship only about ten per cent shorter than the Stella Lykes and with the same beam and displacement. Our chief mate, J. Peter Fritz, wished he were aboard her. She was headed for Narragansett Bay, her regular run, and his home is on Narragansett Bay. He grew up there. As a kid, he used to go around on his bike visiting ships. He took photographs aboard the ships, developed and printed them at home, and went back with the pictures to show the crews. They invited him to stay aboard for dinner. ("Some guys liked air-

planes. To me it was just the ships.") He watched the shipping card in the Providence *Journal* — the column that reports arrivals and departures. Working on a tug and barge, he learned basic seamanship from the harbormaster of Pawtuxet Cove — knots, splicing, "how to lay around boats the right way." As a Christmas present an aunt gave him a picture book of merchant ships. As a birthday present she gave him the "American Merchant Seamans Manual."

Peter grew up, graduated from the Massachusetts Maritime Academy, and went to sea. It was his calling, and he loved it. He also loved, seriatim, half the young women in Rhode

Seaman chipping rust on
the deck of the S.S. Export
Commerce, *North Atlantic
Ocean.*

Island. He was a tall, blond warrior out of "The Twilight of the Gods" with an attractively staccato manner of speech. Not even his physical attractions, however, could secure his romantic hatches. "Dear Peter" letters poured in after he left his women and went off for months at sea. Eventually, he married, had a son, and left the Merchant Marine. For several years, he worked for an electronic-alarm company and miserably longed for the ocean. ("I will not admit how much I love this job. The simple life. Having one boss. Not standing still, not being stagnant; the idea of moving, the constant change.") Eventually, he couldn't stand it any longer, and went off to circle the world on the container ship President Harrison. ("I had *the* killer card. I had planned it.") He made more money in eighty-seven days

than he could make in a year ashore. After a family conference, he decided to ship out again. Like his lifelong friend Clayton Babineau, he coveted a job that would take him on short runs from his home port. Every ten days, the Marine Electric went out of Providence for Hampton Roads, and nine days later she came back. She went right past Peter's house. He night-mated her. His wife, Nancy, said to him, "Hey, wouldn't it be great if you got a job on that one? You could be home with your family." He tried repeatedly, without success. His friend Clay Babineau, sailing as second mate, died of hypothermia that night off Chincoteague in the winter storm. The Marine Electric was thirty-nine years old in the bow and stern, younger in the middle, where she had been stretched for bulk cargo. In the language of the

Coast Guard's Marine Casualty Report, her forward hatch covers were "wasted, holed, deteriorated, epoxy patched." Winds were gusting at sixty miles an hour, and the crests of waves were forty feet high. As the Marine Electric plowed the sea, water fell through the hatch covers as if they were colanders. By 1 A.M., the bow was sluggish. Green seas began pouring over it. A list developed. The captain notified the Coast Guard that he had decided to abandon ship. The crew of thirty-four was collecting on the starboard boat deck, but before a lifeboat could be lowered the ship capsized, and the men, in their life jackets, were in the frigid water. In two predawn hours, all but three of them died, while their ship went to rest on the bottom, a hundred and twenty feet below, destroyed by what the Coast Guard called "the dynamic effects of the striking sea."

Peter Fritz, who gives the routine lectures on survival suits to the successive crews of the Stella Lykes, carries in his wallet a shipping card clipped from the February 13, 1983, Providence *Journal*: "ARRIVING TODAY, MARINE ELECTRIC, 8 P.M."

She is remembered as "a rotten ship." So is the Panoceanic Faith, which went out of San Francisco bound for India with a load of fertilizer about six months after Fritz graduated from Massachusetts Maritime. Five of his classmates were aboard, and all of them died, including his friend John McPhee. Getting to know Fritz has not been easy. There have been times when I felt that he regarded me as a black cat that walked under a ladder and up the gangway, a shipmate in a white sheet, a G. A. C. (Ghost in Addition to Crew). The Panoceanic Faith developed a leak, its dampened cargo expanded, its plates cracked. It sank in daytime. "People tried to make it to the life rafts but the cold water got them first."

Plaques at the maritime academies list graduates who have been lost at sea. A schoolmate of Andy Chase was on a ship called Poet that went out of Cape Henlopen in the fall of 1980 with a load of corn. She was never heard from again. Nothing is known. In Captain Washburn's words, "Never found a life jacket, never found a stick."

On the Spray, Andy went through one hurricane three times. A thousand-pound piece of steel pipe broke its lashings and "became the proverbial loose cannon." Ten crewmen — five on a side — held on to a line and eventually managed to control it, but they had almost no sleep for two days. The Spray once carried forty men. Reduced manning had cut the number to twenty. "Companies are trying to get it down to eleven or twelve by automating most functions," he says. "When everything's going right, four people can run a ship, but all the automation in the world can't handle emergencies like that."

A small ship can be destroyed by icing. Ocean spray freezes and thickens on her decks and superstructure. Freezing rain may add to the accumulation. The amount of ice becomes so heavy that the ship almost disappears within it before the toppling weight rolls her over and sinks her. Ships carry baseball bats. Crewmen club the ice, which can thicken an inch an hour.

To riffle through a stack of the *Mariners Weather Log* — a dozen or so quarterly issues — is to develop a stop-action picture of casualties on the sea, of which there are so many hundreds that the eye skips. The story can be taken up and dropped anywhere, with differing names and the same situation unending. You see the Arctic Viking hit an iceberg off Labrador, the Panbali Kamara capsize off Sierra Leone, the Maria Ramos sink off southern Brazil. A ferry with a thousand passengers hits a freighter with a radioactive cargo and sinks her in a Channel fog. A cargo shifts in high winds and the Islamar Tercero goes down with twenty-six, somewhere south of the Canaries. Within a few days of one another, the Dawn Warbler goes aground, the Neyland goes aground, the Lubeca goes aground, the Transporter II throws twenty-six containers, and the Heather Valley — hit by three waves — sinks off western

Scotland. The Chien Chung sinks with twenty-one in high seas east of Brazil, and after two ships collide off Argentina suddenly there is one. A tanker runs ashore in Palm Beach, goes right up on the sand. The bow noses into someone's villa and ends up in the swimming pool. The Nomada, hit by lightning, sinks off Indonesia. The Australian Highway rescues the Nomada's crew. The Blue Angel, with a crew of twenty, sinks in the Philippine Sea. The Golden Pine, with a shifting cargo of logs (what else?) sinks in the Philippine Sea. A hundred and fifteen people on the Asunció drown as she sinks in the South China Sea. The Glenda capsizes off Mindanao, and seven of the twenty-seven are rescued. The Sofia sinks in rough water near Crete, abandoned by her crew. The Arco Anchorage grounds in fog. On the Arco Prudhoe Bay, bound for Valdez, a spare propeller gets loose on the deck and hurtles around smashing pipes. The Vennas, with sixty-nine passengers and crew, sinks in the Celebes Sea. The Castillo de Salas, a bulk carrier with a hundred thousand tons of coal, breaks in two in the Bay of Biscay. The container ship Tuxpan disappears at noon in the middle of the North Atlantic with twenty-seven Mexicans aboard. A container from *inside* the hold is found on the surface. Apparently, the ship was crushed by a wave. In the same storm in the same sea, a wave hits the Export Patriot hard enough to buckle her doors. Water pours into the wheelhouse. The quartermaster is lashed to a bulkhead so that he can steer the ship. In the same storm, the Balsa 24 capsizes with a crew of nineteen. In the Gulf of Mexico, off the mouth of the Rio Grande, fifteen Mexican shark-fishing vessels sink in one squall. In a fog near the entrance to the Baltic Sea, the Swedish freighter Syderfjord is cut in two in a collision and sinks in

forty seconds. About a hundred miles off South Africa, the Arctic Career leaves an oil slick, some scattered debris, and no other clues. . . . The Tina, a bulk carrier under the Cypriot flag, vanishes without a trace somewhere in the Sulu Sea. In a fog in the Formosa Strait, the Quatsino Sound goes down after colliding with the Ever Linking. In the English Channel, the Herald of Free Enterprise overturns with a loss of two hundred. The Soviet freighter Komsomolets Kirgizzii sinks off New Jersey. In the North Sea, the bridge of the St. Sunnivar is smashed by a hundred-foot wave. After a shift of cargo, the Haitian freighter Aristeo capsizes off Florida. On the Queen Elizabeth 2, Captain Lawrence Portet ties himself to a chair on the bridge. Among the eighteen hundred passengers, many bones are broken. Seas approach forty feet. After a series of deep rolls, there are crewmen who admit to fearing she would not come up. Off the Kentish coast with a hundred and thirty-seven thousand tons of

crude, the tanker Skyron, of Liberian registry, plows a Polish freighter. The tanker bursts into flames. The fire is put out before it can reach the crude. . . . Somewhere, any time, something is getting it. . . .

In Peter Fritz's letters home he avoids mentioning storms. He doesn't want to worry Nancy. On his long vacations, as he leans back, stretches his legs, and watches the evening news, a remark by a television reporter will sometimes cause him to sit straight up. To Peter it is the sort of remark that underscores the separateness of the American people from their Merchant Marine, and it makes him feel outcast and lonely. After describing the havoc brought by some weather system to the towns and cities of New England — the number of people left dead — the reporter announces that the danger has passed, for "the storm went safely out to sea."

— JOHN MCPHEE, *Looking for a Ship,* 1990

Albatross IV *in heavy seas on Georges Bank off Massachusetts, 1992. The 187-foot research vessel keeps track of fish stocks in these intensively fished waters.*

LEGENDS AND CUSTOMS OF THE SEA

➤ Scottish law once required fishermen to wear a gold earring, which was used to pay for funeral expenses if they were drowned and washed ashore.

➤ An old custom dictates that any sailor who sails around Cape Horn is entitled to a small blue tattoo in the shape of a five-pointed star on his left ear. Five times around earns a star on the right ear as well, and two red stars on the forehead is the sign of a great voyager who has rounded the Cape ten times or more. According to one sailor, who himself sports a star on his left ear, there are only two red-star men in the world. Both live in Liverpool, where no pub would charge a red-star man for a drink.

➤ Wine poured upon the deck before a long voyage represents a libation to the gods which will bring good luck. "Christening" a ship by breaking a bottle of champagne across her bow at the time of launching arose from this belief.

➤ It was in the early days of the British Navy that guns were first fired in salute. Since they could not be reloaded quickly, the act of firing a gun in salute assured those receiving the salute that those who fired had disarmed themselves, and could do no harm. The more guns that were fired, the greater the assurance of disarmament, and the higher the respect offered to those being saluted. The largest ships of the fleet held twenty-one guns along one side, therefore the highest mark of respect was a twenty-one-gun salute.

➤ During World War II the United States Navy instituted a system for naming various classes of ships, including the following: *Ammunition ships:* for volcanos or names suggesting fire and explosives; *Battleships:* after states of the union; *Destroyers:* in honor of dead persons associated with the Navy or Marines; *Hospital ships:* with "synonyms for kindness" or "other logical and euphonious words"; *Ocean tugs:* for Indian tribes; *Provision storeships:* for astronomical bodies; *Submarines:* after fish and other sea life.

➤ "What the sea wants, the sea will have," according to the traditional wisdom of the British Isles and many maritime cultures. Thus fatalistic sailors of the past — and some of the present — never learned to swim.

➤ Legend has it that an umbrella aboard ship is unlucky.

Christening ceremony to launch a steam yacht, New England, late nineteenth century.

DISCOVERY

DISCOVERY

PREVIOUS SPREAD
Many marine animals remain to be discovered. This giant purple jellyfish, a relative of corals and anemones in the genus Chrysaora, was found in 1990, and the species is new to science. Scientists estimate that between 500,000 and 5,000,000 new species may be found in the seas. Jellyfish living in the open oceans can be enormous, with bodies over eight feet across and tentacles trailing for one hundred thirty feet.

On a cold and blustery autumn night more than a decade ago, I sat in the stern of a small research vessel and gazed down into the black water of the abyssal deep. We were a dozen miles out from the scarp of the extinct volcano that is the foundation of the Bermuda archi-pelago, floating a thousand fathoms above the craggy bottom. No lights were visible anywhere on the sea around us, and I felt alone and vulnerable, comforted only by the faint loom from the invisible islands.

Steel cables angled downward into the darkness from two winch drums mounted on the bulwarks. The first 2,500 feet of cable were bare; the last 500 we had festooned with chemical lights and baited hooks.

Huddled against the wind in the shelter of a bulkhead, warmed by black coffee and a steaming stew of mashed potatoes and boiled hog snout, I sat and waited, half hoping to feel the boat lurch suddenly, from the bite of an unseen beast half a mile below, half hoping for a night of peaceful failure.

The odds favored failure: no one had ever caught a full-sized specimen of the creature we were seeking; few had ever seen one alive.

We were fishing for giant squid — *Architeuthis dux* — one of the last great monsters on earth, and a living nightmare: ten writhing arms studded with chitinous sucker disks, gigantic eyes, a huge, razor-sharp beak that flays prey and jams shredded flesh back toward a rasp-studded tongue.

Giant squid have been sighted for millennia: Homer's person-ification of Scylla in the *Odyssey* is thought to be a portrait of *Architeuthis*; Herman Melville saw one during his sailing days, and recorded it in *Moby-Dick*. They have been documented at more than 50 feet long, and extrapolated (from partial carcasses washed up on beaches) at 75 or 80 feet. Some scientists have speculated, off the record, that a 100-foot giant squid is a distinct possibility. They have been known to attack small boats and life rafts, and have been widely (if unreliably) reported to have assaulted ocean-going ships.

We had tried to prepare ourselves for success. We had brought knives aboard, and hatchets, and side-cutters to sever the cables in case the squid that took our bait was too big to battle. Still, the atmosphere on the boat was charged with tension.

Suppose . . .

What if . . . ?

Dawn came, and with it disappointment. Nothing had happened. There had been no yawing of the stern of the boat, no squeal of stretched cable, no sign whatever of action. Red-eyed and weary, we began to reel in the cables . . .

. . . and found that they had been sheared off. Both of them. At about 2,000 feet. Not merely broken, as if they had somehow caught in the bottom or foul-hooked a whale, for the strands would have popped and curled; not abraded, as they would have been by the grinding teeth of a deep-dwelling shark.

Forty-eight woven strands of stainless steel had been snipped as cleanly as if by sharpened bolt cutters.

Bitten off.

The hair on my neck stood on end, and a shiver scampered up my spine.

I looked at Teddy Tucker, the legendary Bermudian who has seen more in his half-century on the sea than most have dreamed of, and said, "What could have done that?"

"Who knows?" he replied with a smile. "Maybe a squid, maybe not. Maybe something we've never heard of and can't even imagine." He started the boat and turned toward shore. "The sea's a tease, and she always will be. That's what keeps us looking."

Our most sophisticated methods for studying the sea have been likened to trying to learn about life on land by towing a butterfly net behind an airplane. And the facts bear out the simile: it has been less than three decades since man first set foot on the moon, yet we already know more about the surface of the moon than we do about the bottom of our own oceans. An estimated 95 percent of the sea floor remains to be mapped in detail.

Small wonder, if you consider the challenge: the oceans that blanket the planet in perpetual darkness hide mountains higher than the Himalayas, valleys deeper than the Grand Canyon, and plains vaster than the Gobi Desert — all squeezed by weight and pressure so enormous that they can crush a dreadnought to rubble and reduce terrestrial life to jelly in the blink of an eye.

Compared to the task of exploring the deep sea, scaling the summit of Mount Everest is a stroll in the park.

I well recall the thrill I felt diving to 200 feet for the first time, awed by the profusion of exotic life, frightened by the enveloping darkness — and then suddenly thunderstruck by the knowledge that here, near the limits of safe diving, I had barely scratched the skin of the sea.

In the last 50 years, humans have descended more than 150 times deeper than I had gone that day. They have taken appalling risks and endured incalculable dangers, all in pursuit of the unknown.

I wondered why. What combination of daring and genius, courage and conviction, impels otherwise rational human beings to risk their lives attempting the manifestly impossible?

What must William Beebe have felt when, in 1934, he allowed himself to be lowered 3,028 feet beneath the surface in a tiny steel ball? I know that I, surrounded by pressure of more than half a ton per square inch, might well have babbled in hysterics; Beebe maintained scientific detachment and astonishing cool. "We had no realization of the outside pressure," he wrote, "but the blackness itself seemed to close in on us."

What kind of courage did it take for Jacques-Yves Cousteau, one morning in 1943, to walk into the Mediterranean and, wearing a tank of compressed air and an experimental breathing hose, submerge himself into a medium 800 times denser than air? Imagine the sensation of delight for the Smithsonian Institution's Dr. James Mead when, in 1991, he startled the world with the announcement of his discovery of an entirely new species of whale: *Mesoplodon peruvianus*.

Of course, giant leaps in technology have increased exponentially our ability to explore the undersea world. Without the development of submersibles and ROVs (Remote Operated Vehicles), scientists would never have been able to visit and photograph the wreck of the *Titanic*, which had lain hidden beneath more than two miles of ocean since 1912.

Nor would they have been able to make their (literally) earth-shaking discovery of hydrothermal vents on the deep-sea floor — cones that spew a superheated smokelike fluid from the bowels of the earth.

As important as submersibles are, however, they have as yet been able to afford us only the most fleeting glimpse of all that lies down there in the darkness. I like James Hamilton-Paterson's analogy in *The Great Deep*:

"One might compare [exploring by submersible] to traveling across Asia by oil-lit hansom cab with the conditions of a Dickensian fog outside, and then claiming to have seen the world."

The newest allies of ocean science are satellites, which can accomplish Herculean tasks in a fraction of the time required by traditional methods. A satellite can take thorough and accurate measurements of vitually the entire planet in ten days.

Still, whenever I read dazzling reports of technological breakthroughs, I remind myself that as the twenty-first century approaches, we have acquainted ourselves with only 5 percent of the oceans of our world. And when we talk of the sea, we find ourselves confronted with many more questions than answers.

As for the giant squid, I have continued to look for him every year. I have dived with sperm whales, which feed on him. I have sent lights and cameras and tantalizing baits into the abyss, hoping to lure him to my sight.

I have never found him.

I will keep searching, but, in a way, I hope I never find him, for he is one of our last dragons, and I believe deeply that we need dragons, to keep alive our sense of mystery and adventure.

To me, the giant squid — beautiful, horrible, efficient, elusive, terrifying, mythical, and real — represents the frontier of the sea, still out there, still beckoning.

— PETER BENCHLEY

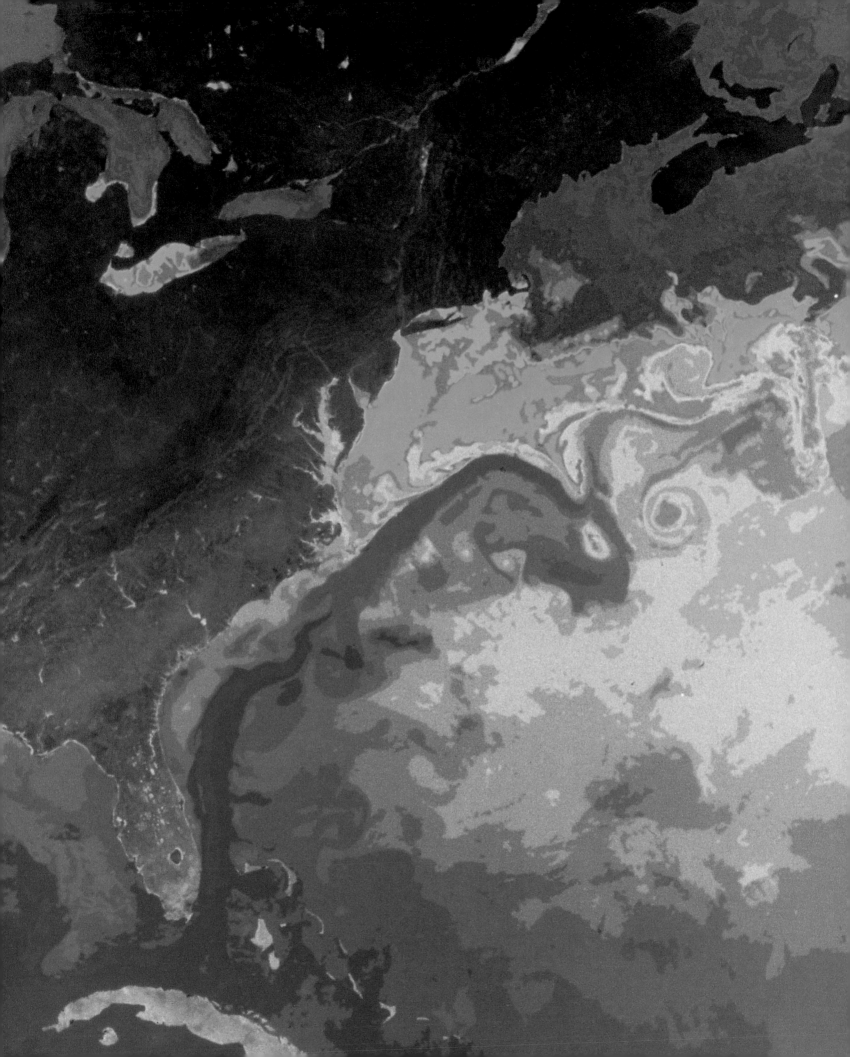

ON THE GULF STREAM

THANKS TO MODERN SATELLITE TECHNOLOGY, SCIENCE IS GAINING A FULLER UNDERSTANDING OF THE INFLUENCE OF MAJOR OCEAN CURRENTS ON EVERYTHING FROM WEATHER TO SHIPPING TO FISHING. TWO HUNDRED YEARS AGO, HOWEVER, OCEAN CURRENTS WERE REGARDED MORE AS MYSTERIES THAN MIRACLES. IT TOOK A GENIUS NAMED BENJAMIN FRANKLIN, THEN DEPUTY POSTMASTER GENERAL OF THE BRITISH COLONIES IN NORTH AMERICA TRYING TO SPEED UP THE MAIL, WORKING WITH AN EXPERIENCED NANTUCKET WHALING CAPTAIN NAMED TIMOTHY FOLGER, TO UNRAVEL THE ENIGMA OF THE GULF STREAM. NOW WE KNOW THAT HEAT FROM THE GULF STREAM, WHICH FLOWS NORTH ALONG THE EAST COAST OF THE UNITED STATES, AND WESTERLY WINDS ARE CRITICAL TO MAINTAINING WESTERN EUROPE'S MODERATE CLIMATE.

A satellite produced this image of sea surface temperatures in the western North Atlantic. Warmer water is yellow, and the warm Gulf Stream is clearly visible as it flows up the east coast of the United States and then across the Atlantic. This 1984 image is surprisingly similar to the map produced by Benjamin Franklin over two hundred years ago.

Vessels are sometimes retarded, and sometimes forwarded in their voyages, by currents at sea, which are often not perceived. About the year 1769 or 70, there was an application made by the board of customs at Boston, to the lords of the treasury in London, complaining that the packets between Falmouth and New-York, were generally a fortnight longer in their passages, than merchant ships from London to Rhode-Island, and proposing that for the future they should be ordered to Rhode-Island instead of New-York. Being then concerned in the management of the American post-office, I happened to be consulted on the occasion; and it appearing strange to me that there should be such a difference between two places, scarce a day's run asunder, especially when the merchant ships are generally deeper laden, and more weakly manned than the packets, and had from London the whole length of the river and channel to run before they left the land of England, while the packets had only to go from Falmouth, I could not but think the fact misunderstood or misrepresented. There happened then to be in London, a Nantucket sea-captain of my acquaintance, to whom I communicated the affair. He told me he believed the fact might be true; but the difference was owing to this, that the Rhode-Island captains were acquainted with the gulf stream, which those of English packets were not. We are well acquainted with that stream, says he, because in our pursuit of whales, which keep near the sides of it, but are not to be met with in it, we run down along the sides, and frequently cross it to change our side: and in crossing it have sometimes met and spoke with those packets, who were in the middle of it, and stemming it. We have informed them that they were stemming a current, that was against them to the value of three miles an hour; and advised them to cross it and get out of it; but they were too wise to be counselled by simple American fishermen. When the winds are but light, he added, they are carried back by the current more than they are forwarded by the wind: and if the wind be good, the subtraction of 70 miles a day from their course is of some importance. I then observed that it was a pity no notice was taken of this current upon the charts, and requested him to mark it out for me, which he readily complied with, adding directions for avoiding it in sailing from Europe to North-America. I procured it to be engraved by order from the general post-office, on the old chart of the Atlantic, at Mount and Page's, Tower-hill; and copies were sent down to Falmouth for the captains of the packets, who slighted it however; but it is since printed in France, of which edition I hereto annex a copy.

This stream is probably generated by the great accumulation of water on the eastern coast of America between the tropics, by the trade winds which constantly blow there. It is known that a large piece of water ten miles broad and generally only three feet deep, has by a strong wind had its waters driven to one side and sustained so as to become six feet deep, while the windward side was laid dry. This may give some idea of the quantity heaped up on the American coast, and the reason of its running down in a strong current through the islands into the bay of Mexico, and from thence issuing through the gulph of Florida, and proceeding along the coast to the banks of Newfoundland, where it turns off towards and runs down through the Western islands. Having since crossed this stream several times in passing between America and Europe, I have been attentive to sundry circumstances relating to it, by which it is interspersed. I find that it is always warmer than the sea on each side of it, and that it does not sparkle in the night: I annex hereto the observations made with the thermometer in two voyages, and possibly may add a third. It will appear from them, that the thermometer may be an useful instrument to a navigator, since currents coming from the northward into southern seas, will probably be found colder than the water of those seas, as the currents from southern seas into northern are found warmer. And it is not to be wondered that so vast a body of deep warm water, several leagues wide, coming from between the tropics and issuing out of the gulph into the northern seas, should retain its warmth longer than the twenty or thirty days required to its passing the banks of Newfoundland. The quantity is too great, and it is too deep to be suddenly cooled by passing under a cooler air. The air immediately over it, however, may receive so much warmth from it as to be rarified and rise, being rendered lighter than the air on each side of the stream; hence those airs must flow in to supply the place of the rising warm air, and meeting with each other, form those tornados and warm-spouts frequently met with, and seen near and over the stream; and as the vapour from a cup of tea in a warm room, and the breath of an animal in the same room, are hardly visible, but become sensible immediately when out in the cold air, so the vapour from the gulph stream, in warm latitudes is scarcely visible, but when it comes into the cool air from Newfoundland, it is condensed into the fogs, for which those parts are so remarkable.

The power of wind to raise water above its common level in the sea, is known to us in America, by the high tides occasioned in all our sea-ports when a strong northeaster blows against the gulph stream.

The conclusion from these remarks is, that a vessel from Europe to North-America may shorten her passage by avoiding to stem the stream, in which the thermometer will be very useful; and a vessel from America to Europe may do the same by the same means of keeping in it. It may often happen accidentally, that voyages have been shortened by these circumstances. It is well to have the command of them.

— BENJAMIN FRANKLIN, 1786

Franklin and Folger had maps showing the Gulf Stream engraved in 1769 to help speed the mail from Europe. Systematic study of the Gulf Stream was begun in the nineteenth-century by Matthew Fontaine Maury and A.D. Bache (Franklin's great-grandson).

Ocean waters are constantly
on the move. On May 27,
1990, a storm-tossed cargo
ship spilled 60,000 Nikes in
the north Pacific, and they
drifted on a current known
as the North Pacific Gyre
for at least four years. These
shoes from the spill were
collected on beaches in the
Pacific Northwest at the
locations written on the soles.
Oceanographers tracked the
progress of the shoes, using
the data to confirm a computer model of Pacific ocean
currents.

A DARK AND LUMINOUS BLUE

WE HAVE SEEN THE TITANIC LYING IN HER GRAVE MORE THAN 12,000 FEET BENEATH THE SURFACE OF THE NORTH ATLANTIC; WE HAVE SEEN THE CRABS AND TUBE WORMS AND OTHER UNLIKELY CREATURES THAT INHABIT CRACKS IN THE SEA FLOOR FAR BEYOND THE REACH OF LIGHT; WE HAVE EVEN SEEN THE CHALLENGER DEEP, THE MOST REMOTE OF ALL THE PLACES ON EARTH. SUBMERSIBLES, ROBOTS, AND VIDEO CAMERAS HAVE ROUTINELY GIVEN US ACCESS TO THE MOST EXOTIC SPOTS IN THE SEA. BUT WHAT WAS IT LIKE TO BE THE FIRST HUMANS TO VENTURE INTO THE DEEP? IN 1934, WILLIAM BEEBE, A NATURALIST, AND OTIS BARTON, AN ENGINEER, CLIMBED INTO A CRUDE STEEL BALL CALLED THE BATHSPHERE AND DESCENDED HALF A MILE INTO THE UNKNOWN DARKNESS OFF BERMUDA. DURING THE DIVES, BEEBE COMMUNICATED WITH HIS COLLEAGUE, MISS HOLLISTER, WHO WAS ON THE SURFACE, VIA TELEPHONE.

William Beebe and the bathysphere, the first serious deep submergence chamber. Beebe and the vessel's builder, Otis Barton, crouched for hours in the frigid steel sphere peering through portholes made of 3-inch-thick fused quartz. Their record-breaking dives in the 1930s and Beebe's lucid observations opened up the world of the deep sea.

Adequate presentation of what I saw on this dive is one of the most difficult things I ever attempted. It corresponds precisely to putting the question, "What do you think of America?" to a foreigner who has spent a few hours in New York City. Only the five of us who have gone down even to 1000 feet in the bathysphere know how hard it is to find words to translate this alien world.

This dive turned out to be one of essential observation, and first hand impressions must take precedence over all others.

At 9:41 in the morning we splashed beneath the surface, and often as I have experienced it, the sudden shift from a golden yellow world to a green one was unexpected. After the foam and bubbles passed from the glass, we were bathed in green; our faces, the tanks, the trays, even the blackened walls were tinged. . . . The first plunge erases, to the eye, all the comforting, warm rays of the spectrum. The red and the orange are as if they had never been, and soon the yellow is swallowed up in the green. We cherish all these on the surface of the earth and when they are winnowed out at 100 feet or more, although they are only one-sixth of the visible spectrum, yet, in our mind, all the rest belongs to chill and night and death. Even modern war bears this out; no more are red blood and scarlet flames its symbols, but the terrible grayness of gas, the ghastly blue of Very lights. . . .

At 320 feet a lovely colony of siphonophores drifted past. At this level they appeared like spun glass. Others which I saw at far greater and blacker depths were illumined, but whether by their own or by reflected light I cannot say. These are colonial creatures like submerged Portuguese men-o'-war, and similar to those beautiful beings are composed of a colony of individuals, which perform separate functions, such as flotation, swimming, stinging, feeding, and breeding, all joined by the common bond of a food canal. Here in their own haunts they swept slowly along like an inverted spray of lilies-of-the-valley, alive and in constant motion. In our nets we find only the half-broken swimming bells, like cracked, crystal chalices, with all the wonderful loops and tendrils and animal flowers completely lost or contracted into a mass of tangled threads. Twenty feet lower a pilotfish looked in upon me — the companion of sharks and turtles, which we usually think of as a surface fish, but with only our pitiful, two-dimensional, human observation for proof.

When scores of bathyspheres are in use we shall know much more about the vertical distribution of fish than we do now. For example, my next visitors were good-sized yellow-tails and two blue-banded jacks which examined me closely at 400 and 490 feet respectively. Here were so-called surface fish happy at 80 fathoms. Several silvery squid balanced for a moment, then shot past, and at 500 feet a pair of lanternfish with no lights showing looked at the bathysphere unafraid.

At 600 feet the color appeared to be a dark,

luminous blue, and this contradiction of terms shows the difficulty of description. As in former dives, it seemed bright, but was so lacking in actual power that it was useless for reading and writing. . . .

There are certain nodes of emotion in a descent such as this, the first of which is the initial flash. This came at 670 feet, and it seemed to close a door upon the upper world. Green, the world-wide color of plants, had long since disappeared from our new cosmos, just as the last plants of the sea themselves had been left behind far over head.

At 700 feet the light beam from our bulb was still rather dim; the sun had not given up and was doing his best to assert his power. At 800 feet we passed through a swarm of small beings, copepods, sagitta or arrow worms and every now and then a worm which was not a worm but a fish, one of the innumerable round-mouths or *Cyclothones.* Eighty feet farther and a school of about 30 lanternfish passed, wheeled and returned; I could guess *Myctophum laternatum,* but I cannot be certain. The beam of light drove them away. . . .

Lights now brightened and increased, and at 1100 feet I saw more fish and other organisms than my prebathysphere experience had led me to hope to see on the entire dive. With the light on, several chunky little hatchet-fish approached and passed through; then a silver-eyed larval fish two inches long; a jelly; suddenly a vision to which I can give no name, although I saw others subsequently. It was a network of luminosity, deli-cate, with large meshes, all aglow and in motion, waving slowly as it drifted. Next a dim, very deeply built fish appeared and vanished; then a four-inch larval eel swimming obliquely upward; and so on. This ceaseless telephoning left me breathless and I was glad of a hundred feet of only blue-blackness and active sparks. . . .

Suddenly in the distance a strong glow shot forth, covering a space of perhaps eight inches. Not even the wildest guess would help with such an occurrence. Then the law of compensation

sent, close to the window, a clear-cut, three-inch, black anglerfish with a pale, lemon-colored light on a slender tentacle. All else my eye missed, so I can never give it a name.

One great source of trouble in this bathy-sphere work is the lag of mind behind instanta-neous observation. For example, at 1300 feet a medium-sized, wide-mouthed angler came in sight, then vanished, and I was automatically describing an eight-inch larval eel looking like a transparent willow leaf, when my mind shot back to the angler and demanded how I had seen it. I had recorded no individual lights on body or tentacle, and now I realized that the teeth had glowed dully, the two rows of fangs were luminous. It is most baffling to gaze into outer darkness, suddenly see a vision, record the bare facts — the generality of the thing itself — and then, in the face of complete distraction by another spark or organism, to have to hark back and recall what special characters escaped the mind but were momentarily etched upon the reti-na. On this point I had thoroughly coached Miss Hollister at the other end of the telephone, so I constantly received a fire of questions, which served to focus my attention and flick my memo-ry. Again and again when such a question came, I willfully shut my eyes or turned them into the bathysphere to avoid whatever bewilderment might come while I was searching my memory for details of what had barely faded from my eye. . . .

At 1500 feet I swung for two and a half min-utes, and here occurred the second memorable moment in these dives — opportunity for the deliberate, accurate record of a fish wholly new to science, seen by one or both of us, the proof of whose existence, other than our word, must await the luck of capture in nets far more effective than those we now use in our oceanographic work. First, a quartet of slender, elongate fish passed through the electric light literally like arrows, about twenty inches long, whether eels or not I shall never know; then a jelly, so close that it

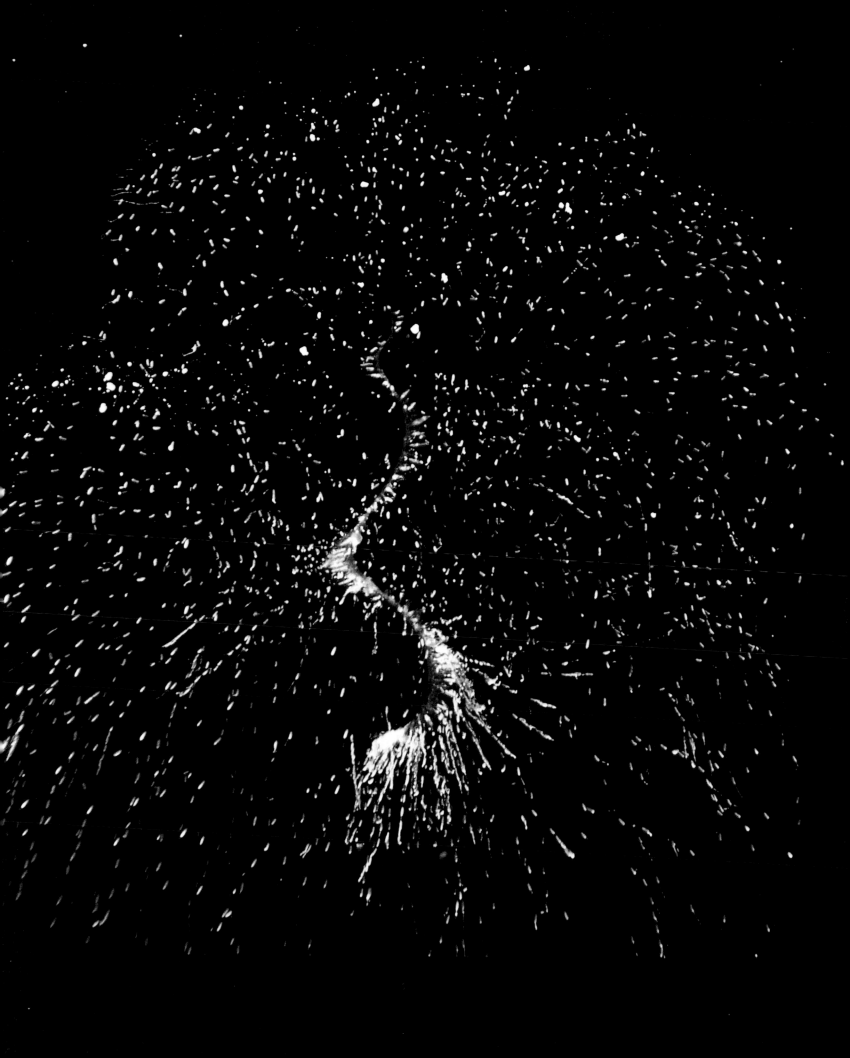

almost brushed the glass. Finally, without my seeing how it got there, a large fish swung suspended, half in, half out of the beam. It was poised with only a slow waving of fins. I saw it was something wholly unknown, and I did two things at once; I reached behind for Mr. Barton, to drag him away from his camera preparations to the windows, to see and corroborate, and I disregarded Miss Hollister's insistent questions in my ears. I had to grunt or say something in reply to her, for I had already exceeded the five seconds which was our danger duration of silence throughout all the dives. But all this time I sat absorbing the fish from head to tail through the wordless, short-circuiting of sight, later to be materialized into spoken and written words, and finally into a painting dictated by what I had seen through the clear quartz. . . .

At 1630 feet a light grew to twice its diameter before our eyes, until it was fully the diameter of a penny, appearing to emanate from some creature which bore irregular patches of dull luminosity on its body. The outline was too indistinct to tell whether it was with or without a backbone.

At 1900 feet, to my surprise, there was still the faintest hint of dead gray light, 200 feet deeper than usual, attesting the almost complete calm of the surface and the extreme brilliancy of the day far overhead. At 2000 feet the world was forever black. And this I count as the third great moment of descent, when the sun, source of all light and heat on the earth, has been left behind. It is only a psychological mile-post, but it is a very real one. We had no realization of the outside pressure but the blackness itself seemed to close in on us.

At 2000 feet I made careful count and found that there were never less than ten or more lights — pale yellow and pale bluish — in sight at any one time. Fifty feet below I saw another pyrotechnic network, this time, at a conservative estimate, covering an extent of two by three feet. I could trace mesh after mesh in the darkness, but could not even hazard a guess at the cause.

It must be some invertebrate form of life, but so delicate and evanescent that its abyssal form is quite lost if ever we take it in our nets. Another hundred feet and Mr. Barton saw two lights blinking on and off, obviously under control of the fish. . . .

At 2300 feet some exclamation of mine was interrupted by a request from above to listen to the tug's whistles saluting our new record, and my response was, "Thanks ever so much, but take this: two very large leptocephali have just passed through the light, close together, vibrating swiftly along; note — why should larval eels go in pairs?" And with this the inhabitants of our dimly remembered upper world gave up their kindly efforts to honor us. On down we went through a rich, light-filled 2400, and to rest at 2500 feet, for a long half hour. . . .

The several nodes of high lights of which I have written occur on every descent, but there is in addition a compounding of sensations. At first we are quick to see every light, facile in sending up notes, but when we have used up most of our adjectives it is difficult to ring changes on sparks, lights, and darkness. More and more complete severance with the upper world follows, and a plunging into new strangenesses, unpredictable sights continually opening up, until our vocabularies are pauperized, and our minds drugged.

Over two hours had passed since we left the deck and I knew that the nerves both of my staff and myself were getting ragged with constant tenseness and strain. My eyes were weary with the flashing of eternal lights, each of which had to be watched so carefully, and my mind was surfeited with visions of the continual succession of fish and other organisms, and alternately encouraged and depressed by the successful or abortive attempts at identification. So I asked for our ascent.

—WILLIAM BEEBE, *Half Mile Down*, 1934

Before submersibles, scientists used nets to study deep-sea marine life, often missing mobile animals and damaging fragile ones. A modern research sub, the Johnson Sea Link *dives in the Bahamas, scouting cliffs and the sea floor down to depths of 3,000 feet. Scientists observe marine life by peering through the acrylic sphere or through video cameras, and collect specimens using vacuum slurpers, mechanical jars, and robot arms.*

DEEP WATERS

DIVING RECORDS

FREE DIVE (JACQUES MAYOL)	282 FEET
SCUBA DIVE (BRENT GILLIAM)	325 FEET
ALGAE	900 FEET
RAY OF LIGHT FROM THE SUN PERCEPTIBLE TO HUMAN EYE	1,600 FEET
MILITARY SUBMARINE	2,950 FEET
BATHYSPHERE	3,000 FEET
JOHNSON SEA LINK (FOUR-PERSON SUBMERSIBLE)	3,000 FEET
SPERM WHALE	3,770 FEET
NORTHERN ELEPHANT SEAL	4,920 FEET
TITANIC'S FINAL RESTING PLACE	12,500 FEET
ALVIN (THREE-PERSON SUBMERSIBLE)	14,770 FEET
OCTOPUS	16,400 FEET
SPONGE	18,500 FEET
JASON (ROBOTIC SUBMARINE)	19,685 FEET
SHINKAI (THREE-PERSON SUBMERSIBLE)	21,325 FEET
FISH	27,460 FEET
AMPHIPOD (CRAB RELATIVE)	32,200 FEET
DEEPEST SPOT ON EARTH (CHALLENGER DEEP)	35,802 FEET

THESE GIANT SINGLE-CELLED RELATIVES OF BACTERIA CALLED XENO-PHYOPHORES, PHOTO-GRAPHED AT 13,000 FEET IN THE ATLANTIC OFF THE NORTHWEST COAST OF AFRICA, GROW TO BE AN INCH ACROSS AND ARE COMMONLY FOUND ON THE BOTTOM OF THE OCEAN. IN SOME AREAS, THE SEA FLOOR IS COV-ERED BY THE SLIME THEY PRODUCE WHILE FEEDING.

BIOLUMINESENCE

SUNLIGHT FADES RAPIDLY DURING ITS
DESCENT INTO THE OCEANS, BECOMING
EXTREMELY FAINT AT ABOUT 650 FEET.
DEEPER THAN THIS FANTASTIC CREATURES
ARE FOUND, MANY OF WHICH GLOW IN
THE DARK.

RIGHT DEEP-SEA
ANGLER FISH USE BIOLU-
MINESCENCE TO LURE
QUARRY WITHIN REACH
OF THEIR GAPING MOUTHS.
A GLOWING BULB AT THE
END OF A MODIFIED FIN-
RAY PROTRUDES OVER
THE FISH'S NEEDLE-SHARP
TEETH AND ATTRACTS A
VARIETY OF PREY RANGING
FROM COPEPODS TO FISH.

OPPOSITE VIEWED IN
THE BAHAMAS AT 2,000
FEET, THIS SQUID HAS A
BIOLUMINESCENT PATTERN
ON ITS UNDER SURFACE.
LIKE A ZEBRA'S STRIPES,
THEY BREAK UP ITS SIL-
HOUETTE, MAKING IT
HARDER FOR PREDATORS
CRUISING BELOW TO SEE
THE SQUID AGAINST DOWN-
WELLING SUNLIGHT OR
MOONLIGHT. THE SQUID
CAN CONTROL ITS BIOLU-
MINESCENCE SO AS TO
MATCH THE LIGHT AROUND
IT IN BOTH COLOR AND
INTENSITY.

➤ Ninety percent of all volcanic activity occurs in the oceans. In 1993, scientists located the largest known concentration of active volcanoes on the sea floor in the South Pacific. This area, the size of New York State, hosts 1,133 volcanic cones and sea mounts. Two or three of these could be erupting at any moment.

➤ The highest tides in the world are at the Bay of Fundy in Canada. At some times of the year the difference between high and low tides is 53 1/2 feet, the equivalent of a three-story building.

➤ The oceans cover 71 percent of the earth's surface and contain 97 percent of the Earth's water. Less than 1 percent is fresh water, and the remaining 2–3 percent is contained in glaciers and ice caps.

➤ Earth's longest mountain range is the Mid-Ocean Ridge, which winds underwater from the Arctic Ocean to the Atlantic, skirting Africa, Asia, and Australia, and crossing the Pacific to the west coast of North America. It is four times longer than the Andes, Rockies, and Himalayas combined.

➤ Canada has the longest coastline of any country, at 56,453 miles. This is around 15 percent of the world's 372,384 miles of coastlines.

➤ A slow cascade of water beneath the Denmark Strait sinks 2.2 miles, over three and a half times farther than Venezuela's Angel Falls, the tallest waterfall on land.

➤ El Niño, a periodic shift of warm waters from the western to eastern Pacific Ocean, has dramatic effects on climate worldwide. In 1982–3, the most severe El Niño of the century created droughts, crop failures, fires, torrential rains, floods, landslides, and total damages were estimated at over $8 billion.

➤ At the deepest point in the ocean the pressure is over eight tons per square inch, or the equivalent of a person supporting about fifty jumbo jets.

➤ The temperature of almost all of the deep ocean is only a few degrees above freezing, 39 degrees Fahrenheit.

➤ In 1958 the United States Coast Guard icebreaker *East Wind* measured the world's tallest known iceberg off of western Greenland. At 550 feet it was only five and a half feet shorter than the Washington Monument.

➤ If the oceans could be dumped off Earth into open space and allowed to freeze into a ball of ice, it would be about one-third the size of our own moon.

➤ Although Mount Everest, at 29,028 feet, is often called the tallest mountain on Earth, Mauna Kea, an inactive volcano on the island of Hawaii is actually taller. Only 13,796 feet of Mauna Kea stands above sea level, yet it is 33,465 feet tall if measured from the ocean floor to its summit.

➤ If mined, all the gold suspended in the world's seawater would give each person on earth nine pounds.

➤ Undersea earthquakes and other disturbances cause tsunamis, or great waves. The largest recorded tsunami was 210 feet above sea level when it reached Siberia's Kamchatka Peninsula in 1737.

Satellites can measure ocean color thanks to the presence of microscopic algae that absorb certain wavelengths of light and reflect others. Areas with abundant sunlight and nutrients provide ideal growing conditions for phytoplankton, shown here in red. These highly "productive" areas attract animals, often including the fish we eat. A satellite collected data for this map of sea surface color around Tasmania in just two minutes. It would have taken a ship moving at ten knots per hour an entire decade to collect the same type of data.

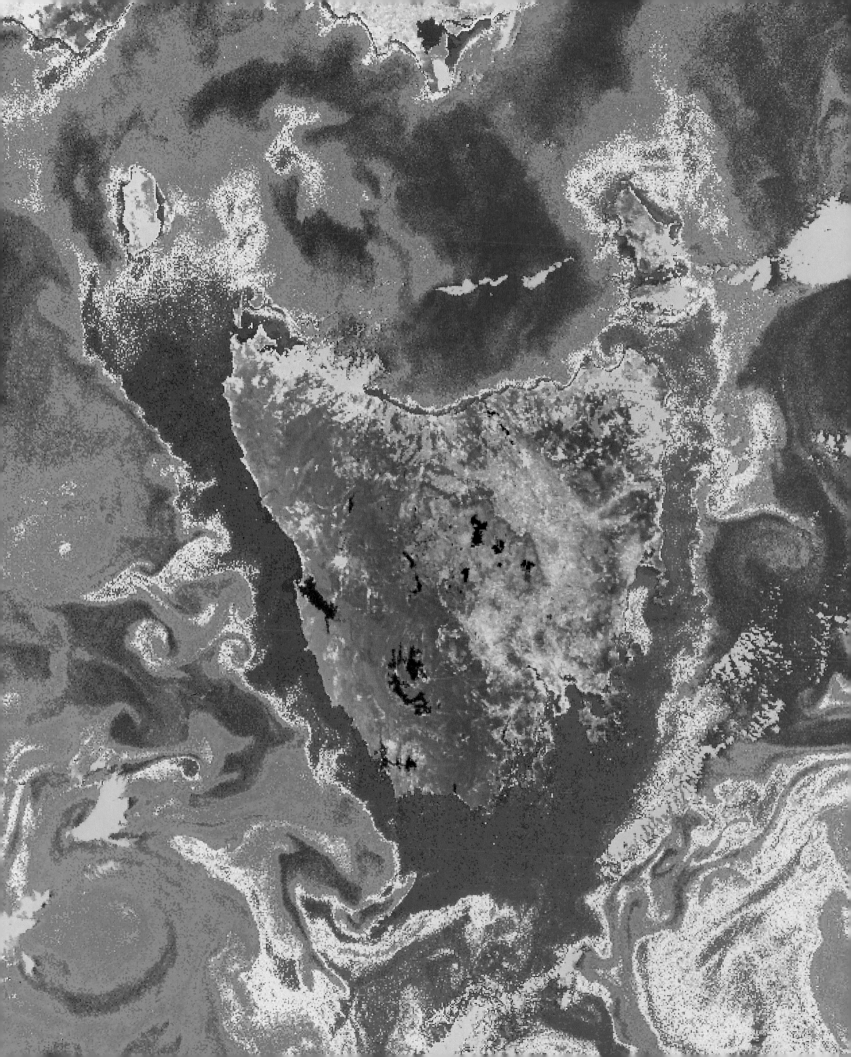

AQUALUNG

THE OCEANS CONTAIN 99 PER-
CENT OF ALL THE LIVING SPACE
ON THE PLANET, BUT UNTIL
HALF A CENTURY AGO, NO MORE
THAN AN INFINITESIMAL FRAC-
TION OF THE WATER WORLD HAD
BEEN EXPLORED. LIMITED BY
OUR NEED TO BREATHE AIR, WE
WERE UNABLE TO DESCEND
MORE THAN A FEW FEET FOR
MORE THAN A FEW MINUTES. IT
SEEMED WE MIGHT REMAIN FOR-
EVER ALIEN TO THE SEA. AND
THEN, AT THE HEIGHT OF WORLD
WAR II, TWO FRENCHMEN,
JACQUES-YVES COUSTEAU AND
EMILE GAGNON, DEVELOPED
AND TESTED A NEW DEVICE THAT
WOULD ENABLE A DIVER TO
CARRY ON HIS BACK HIS OWN
SUPPLY OF COMPRESSED AIR.
THE RESULT WAS A REVOLUTION
IN SCIENCE, DISCOVERY, SPORT,
AND BUSINESS. TODAY, THE
SCUBA-DIVING INDUSTRY GENER-
ATES $2.5 BILLION EVERY YEAR,
AND MILLIONS OF DIVERS ARE
DISCOVERING AND DOCUMENTING
THE WONDERS OF THE LIVING
SEAS.

*The invention of scuba
(self-contained underwater
breathing apparatus) opened
the upper three hundred
feet of the seas to exploration
and recreation.*

One morning in June, 1943, I went to the railway station at Bandol on the French Riviera and received a wooden case expressed from Paris. In it was a new and promising device, the result of years of struggle and dreams, an automatic compressed-air diving lung conceived by Émile Gagnan and myself. I rushed it to Villa Barry where my diving comrades, Philippe Tailliez and Frédéric Dumas waited. No children ever opened a Christmas present with more excitement than ours when we unpacked the first "aqualung." If it worked, diving could be revolutionized.

We found an assembly of three moderate-sized cylinders of compressed air, linked to an air regulator the size of an alarm clock. From the regulator there extended two tubes, joining on a mouthpiece. With this equipment harnessed to the back, a watertight glass mask over the eyes and nose, and rubber foot fins, we intended to make unencumbered flights in the depths of the sea.

We hurried to a sheltered cove which would conceal our activity from curious bathers and Italian occupation troops. I checked the air pressure. The bottles contained air condensed to one hundred and fifty times atmospheric pressure. It was difficult to contain my excitement and discuss calmly the plan of the first dive. Dumas, the best goggle diver in France, would stay on shore keep-ing warm and rested, ready to dive to my aid, if necessary. My wife, Simone, would swim out on the surface with a schnorkel breathing tube and watch me through her submerged mask. If she

signaled anything had gone wrong, Dumas could dive to me in seconds. "Didi," as he was known on the Riviera, could skin dive to sixty feet.

My friends harnessed the three-cylinder block on my back with the regulator riding at the nape of my neck and the hoses looped over my head. I spat on the inside of my shatterproof glass mask and rinsed it in the surf, so that mist would not form inside. I molded the soft rubber flanges of the mask tightly over forehead and cheekbones. I fitted the mouthpiece under my lips and gripped the nodules between my teeth. A vent the size of a paper clip was to pass my inhalations and exhalations beneath the sea. Staggering under the fifty-pound apparatus, I walked with a Charlie Chaplin waddle into the sea. . . .

I looked into the sea with the same sense of trespass that I have felt on every dive. A modest canyon opened below, full of dark green weeds, black sea urchins and small flowerlike white algae. Fingerlings browsed in the scene. The sand sloped down into a clear blue infinity. The sun struck so brightly I had to squint. My arms hanging at my sides, I kicked the fins languidly and traveled down, gaining speed, watching the beach reeling past. I stopped kicking and the momentum carried me on a fabulous glide. When I stopped, I slowly emptied my lungs and held my breath. The diminished volume of my body decreased the lifting force of water, and I sank dreamily down. I inhaled a great chestful and retained it. I rose toward the surface.

My human lungs had a new role to play, that of a sensitive ballasting system. I took normal breaths in a slow rhythm, bowed my head and swam smoothly down to thirty feet. I felt no increasing water pressure, which at that depth is twice that of the surface. The aqualung automatically fed me increased compressed air to meet the new pressure layer. Through the fragile human lung linings this counter-pressure was being transmitted to the blood stream and instantly spread throughout the incompressible body. My brain received no subjective news of the pressure. I was at ease, except for a pain in the middle ear and sinus cavities. I swallowed as one does in a landing airplane to open my Eustachian tubes and healed the pain. . . .

I reached the bottom in a state of transport. A school of silvery sars (goat bream), round and flat as saucers, swam in a rocky chaos. I looked up and saw the surface shining like a defective mirror. In the center of the looking glass was the trim silhouette of Simone, reduced to a doll. I waved.

The doll waved at me.

I became fascinated with my exhalations. The bubbles swelled on the way up through lighter pressure layers, but were peculiarly flattened like mushroom caps by their eager push against the medium. I conceived the importance bubbles were to have for us in the dives to come. As long as air boiled on the surface all was well below. If the bubbles disappeared there would be anxiety, emergency measures, despair. They roared out of the regulator and kept me company. I felt less alone.

I swam across the rocks and compared myself favorably with the sars. To swim fishlike, horizontally, was the logical method in a medium eight hundred times denser than air. To halt and hang attached to nothing, no lines or air pipe to the surface, was a dream. At night I had often had visions of flying by extending my arms as wings. Now I flew without wings. (Since that first aqualung flight, I have never had a dream of flying.)

I thought of the helmet diver arriving where I was on his ponderous boots and struggling to walk a few yards, obsessed with his umbilici and his head imprisoned in copper. On skin dives I had seen him leaning dangerously forward to make a step, clamped in heavier pressure at the ankles than the head, a cripple in an alien land. From this day forward we would swim across miles of country no man had known, free and level, with our flesh feeling what the fish scales know.

I experimented with all possible maneuvers of the aqualung — loops, somersaults and barrel rolls. I stood upside down on one finger and burst out laughing, a shrill distorted laugh. Nothing I did altered the automatic rhythm of air. Delivered from gravity and buoyancy I flew around in space. . . .

Fifteen minutes had passed since I left the little cove. The regulator lisped in a steady cadence in the ten-fathom layer and I could spend an hour there on my air supply. I determined to stay as long as I could stand the chill. Here were tantalizing crevices we had been obliged to pass fleetingly before. I swam inch-by-inch into a dark narrow tunnel, scraping my chest on the floor and ringing the air tanks on the ceiling. In such situations a man is of two minds. One urges him on toward mystery and the other reminds him that he is a creature with good sense that can keep him alive, if he will use it. I bounced against the ceiling. I'd used one-third of my air and was getting lighter. My brain complained that this foolishness might sever my air hoses. I turned over and hung on my back.

The roof of the cave was thronged with lobsters. They stood there like great flies on a ceiling. Their heads and antennae were pointed toward the cave entrance. I breathed lesser lungsful to keep my chest from touching them. Above water was occupied, ill-fed France. I thought of the hundreds of calories a diver loses in cold water. I selected a pair of one-pound lobsters and carefully plucked them from the roof, without touching their stinging spines. I carried them toward the surface.

Simone had been floating, watching my bubbles wherever I went. She swam down toward me. I handed her the lobsters and went down again as she surfaced. She came up under a rock which bore a torpid Provençal citizen with a fishing pole. He saw a blonde girl emerge from the combers with lobsters wriggling in her hands. She said, "Could you please watch these for me?" and put them on the rock. The fisherman dropped his pole.

Simone made five more surface dives to take lobsters from me and carry them to the rock. I surfaced in the cove, out of the fisherman's sight. Simone claimed her lobster swarm. She said, "Keep one for yourself, *monsieur*. They are very easy to catch if you do as I did."

— Jacques-Yves Cousteau with Frédéric Dumas, *The Silent World*, 1953

THE SUNLIT ZONE

SUNLIT WATERS NEAR THE OCEAN'S SURFACE
SUPPORT MASSIVE NUMBERS OF TINY ALGAE,
CALLED PHYTOPLANKTON, THAT CAPTURE
ENERGY FROM SUNLIGHT AND CARBON DIOXIDE
FROM AIR AND SEAWATER. THESE ARE EATEN
BY TINY ANIMALS, WHO ARE IN TURN EATEN
BY LARGER ANIMALS. PLANT-EATERS CONGRE-
GATE IN THE SHALLOW WATERS, ALONG
WITH THE ANIMALS THAT EAT THEM. THE
SHALLOW WATERS THAT BATHE CONTINENTAL
SHELVES COVER ONLY 10 PERCENT OF THE
EARTH'S SURFACE, YET THEY YIELD 90 PER-
CENT OF THE MARKETABLE FISH.

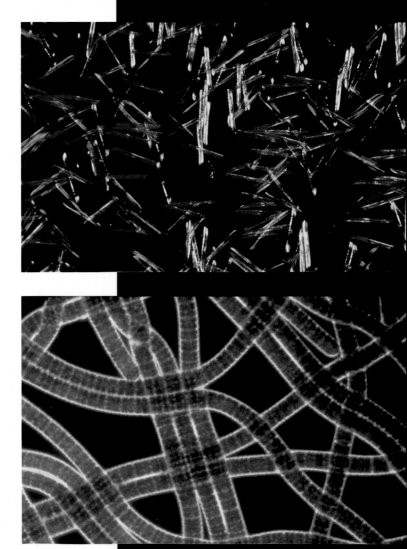

RIGHT DIATOMS ARE
MICROSCOPIC PHOTOSYN-
THETIC ALGAE, OF MANY
SHAPES AND SIZES, WITH
SHELLS OF SILICA. THEY
ARE AMONG THE OCEANS'
MOST NUMEROUS PHOTO-
SYNTHESIZERS AND THE
FIRST AND MOST VITAL
LINK IN FOOD WEBS.

BELOW BLUE-GREEN
ALGAE ARE AMONG THE
OLDEST LIFE FORMS ON
THE PLANET. RELATIVES
OF THESE PHOTOSYNTHE-
SIZING BACTERIAL CELLS
MAY HAVE RELEASED
THE OXYGEN INTO THE
EARTH'S PRIMORDIAL
ATMOSPHERE.

OPPOSITE COPEPODS,
TINY MARINE AND FRESH-
WATER CRUSTACEANS,
ARE THE MOST NUMEROUS
OF GRAZERS IN THE
OCEANS, EARNING THEM
THE NAME "COWS OF THE
SEA." MOST ARE FILTER
FEEDERS THAT SCOOP
UP DIATOMS AND OTHER
PHYTOPLANKTON.

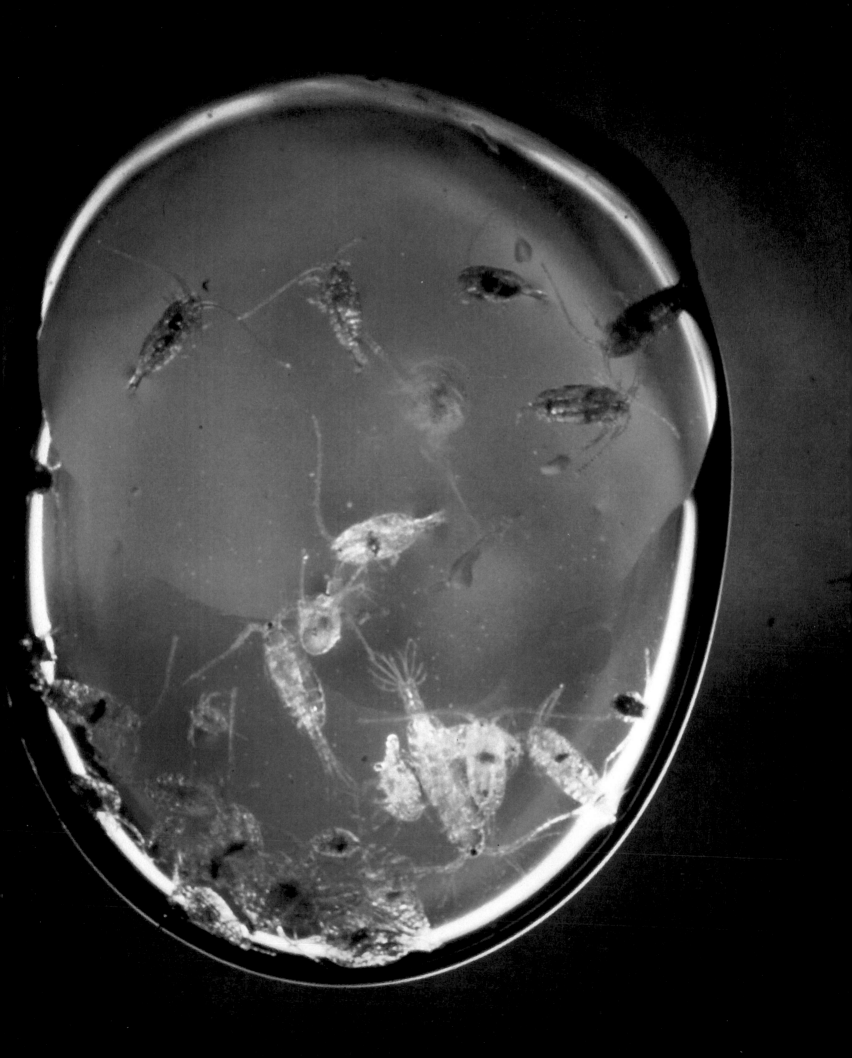

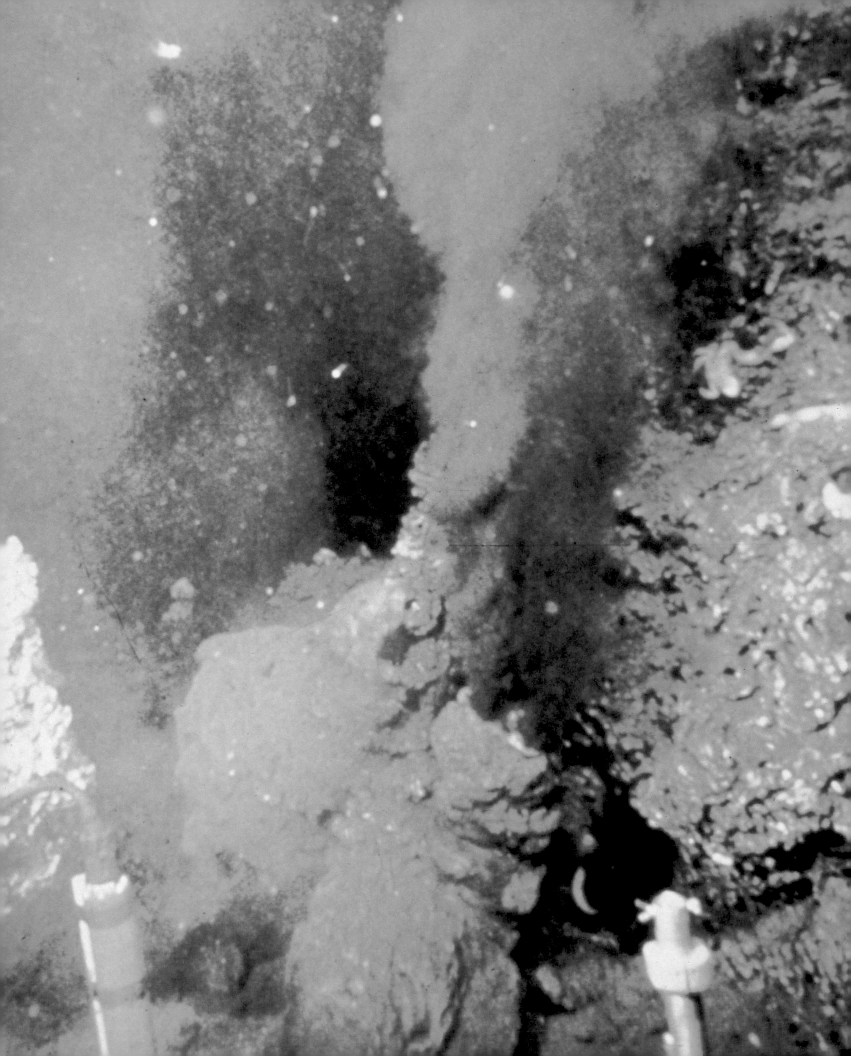

A FIELD OF WORMS

In the late 1970s, scientists studying the movement of the sea floor made an accidental discovery that has since had profound effects on thinking about biology, energy, and even the origin of life on Earth. Nearly two miles deep in the Pacific Ocean, they found dense concentrations of living organisms: worms, bacteria, clams surrounding vents that spewed clouds of hot fluid into the icy water. Reporter Joseph Cone attended this 1984 press conference where Alex Malahoff, a scientist at the National Oceanic and Atmosphereic Administration, announced discoveries made off the coast of Oregon, 9,700 feet down on the Juan de Fuca ridge.

Black smoker photographed from Alvin, *which is visible in the foreground. The hottest submarine hot springs can reach 518–716 degrees Fahrenheit. The superheated water laced with hydrogen sulfide and other minerals spews out of cracks in the Earth's crust.*

When the television crews turned their floodlights on at the back of the conference room, Alex Malahoff, who had just arrived at the podium at the front, squinted. He dropped his eyes from surveying the full room of reporters to glance at his typewritten notes.

He looked up. "Ladies and gentlemen," he began, "this is the culmination of many, many years of effort: to finally go down on the Gorda and Juan de Fuca ridges with a manned submersible."

The scientist looked out over the crowd of reporters. It was a quiet, confident look. More than thirty reporters, representing the mainstream media — the *Los Angeles Times, The New York Times,* Cable News Network — sat expectantly. They were almost all of them far from their newsrooms, and far also, this Saturday morning, from their homes and bedrooms. They had come all the way out to a remote spot on the Pacific coast, to the Marine Science Center of Oregon State University, in hopes of getting an unusual story. But the scientist understood that reporters want something else, too — the feeling that they are not just observers, but in some measure a part of significant events.

He confided in them.

"The ultimate goal in our research is to get the human eye onto the ocean floor," he continued. "Because, although we use sophisticated instrumentation, it is ultimately the human eye and the brain behind it that interprets the phenomena on the ocean floor."

He paused. A New Zealander, he spoke with a slight accent that called attention to his voice. But there was something more to it, something in the way he spoke. It seemed a scientist's habitual manner, straight-ahead, accustomed to stating facts; yet there was an undercurrent, a hint, of the histrionic. Out in the audience were Malahoff's colleagues in the National Oceanic and Atmospheric Administration. Several were wearing little grins of expectation. They had seen Malahoff perform before.

Malahoff continued, setting the stage. He explained that the Gorda and Juan de Fuca ridges were the local continuation of a chain of volcanic mountains on the seafloor that snaked around the globe. The outline of the chain on the earth was like "the strings on a baseball," he said.

The very first exploration of these underwater mountains had begun in the mid-1970s. Several expeditions to submerged mountains in the Atlantic and Pacific oceans had occurred since then. From 1980 the scientists in his team had worked preparing for the dives onto the North Pacific ridges, from which they had just returned. First they had gone to sea to get depth soundings of the ridges. Then, back at their computers, they had turned those depth soundings into maps.

Malahoff turned to a large easel at the side of the podium, where the maps were pinned up. This was the first time they were being shown in public. In shades of red, green, and blue, they

Enormous vent worms — some grow to be almost ten feet long — have no mouth or digestive tract. Instead, chemosynthetic bacteria living in their tissues provide nourishment. Hemoglobin (which carries hydrogen sulfide to the bacteria) makes the worms red.

looked at a distance like striped flags as much as maps: flags to new territories. Malahoff used a pointer to tick off the spots on the map; all eyes in the room followed its movement. Starting in the south on the Gorda Ridge and working up north through the Juan de Fuca, he recounted the dives in the *Alvin*.

The dives on the Gorda had provided many interesting rock and water samples, he said, but had not yielded what the researchers had hoped

to find. They had not seen evidence of volcanic vents on the seafloor.

But then, two researchers — Malahoff himself and Hammond — had dived on a site on the Juan de Fuca, off the Oregon coast, opposite the Columbia River.

"One could just drive out two hundred and fifty miles, if you had a submarine, drop down and see for yourself," Malahoff said, teasing. A few reporters laughed, nervously. The scientist smiled.

"Well," he said, "we found there this rather remarkable scene, if you can imagine it. On the ocean floor, which is normally very cold, you have hot springs — hydrothermal vents.

"They are like oases, and inside these oases you find the most unusual sort of animal communities. We found, for example, five-foot-long worms, hydrothermal worms."

He paused. The room was very quiet. "It is quite extraordinary how you find them," he continued. "They stand in great masses, tall, undulating in the water. They sort of look as though they are part of a field of wheat."

No smile from Malahoff, no underlining of the improbable, bizarre aspects of the scene he was describing. These are facts, please. Poker-facing it.

"When you go past the worms, you come into a hot vent area. This looks like a geyser valley . . . like Yellowstone underwater. But instead of geysers spouting you have chimneys with black smoke coming out of them. The chimneys rise up to heights of forty-five feet and are twisted, grotesquely."

Malahoff pressed on, describing the next dive site, speaking more rapidly. The fluids coming out of the smoking chimneys, he was saying, were at a temperature of about 600 degrees Fahrenheit. "So when you go down into these worm fields, you have to approach very cautiously, because the submarine has a plastic window on the bottom of its hull. You don't want to land on a smoker, and thereby wipe yourself out.

"The visibility is very poor here . . . but what you see is eerily beautiful. The visibility is poor because you have large flocks of bacteria — bacterial clumps the size of snowflakes — floating around."

Abruptly, a small clatter in the room: someone had dropped a pen or pencil. No one turned, no one gave the slightest reaction.

"Underneath this snowstorm of bacteria you see jellylike, translucent, whitish mats on the ocean floor. These are also bacteria."

The reporters could almost be heard struggling mentally to assimilate what they were hearing, to put this information into some familiar context. *Snowing bacteria swirling above a field of worms crowded around underwater hot springs carpeted by more bacteria.*

Malahoff wasn't stopping. "All in all," he was saying, "the life per cubic foot exceeds anything we know of on the surface of the earth." Eyes blinked, almost audibly. "The life density here was just" — he chose his word — "incredible."

He paused then. He wasn't quite done.

"Now in this bag," he said, raising a small plastic bag from the shelf of the podium into view, "I have some specimens."

Immediately, a press photographer who had been standing at the side of the room moved in toward Malahoff, the motor drive on his camera whirring, a second camera around his neck thumping his chest as he walked. At the same time, the reporters in the front row leaned forward to get a better look. People in the back moved forward. Malahoff took clams from the hot spring out of the plastic bag. He took off his glasses. A broad smile transformed his scientist's mask.

When, in 1493, Columbus returned to Spain from his first voyage, he brought with him Indians; and the Spaniards lined the streets to look at these new sorts of beings, naked, red-skinned. Columbus always brought the Indians with him when he visited royalty. They were the token of discoveries that could only, as yet, be talked about. But to king and commoner alike they were signs of something more; they were real, tangible presences to unfetter one's sense of limits.

The reaction of the reporters to the odd little clams was as Malahoff had hoped, because he shared the feeling himself.

This is what anyone wanted from exploration: new worlds to wonder about.

— JOSEPH CONE, *Fire Under the Sea,* 1991

FORMS OF LIFE

LIFE IN A WATERY MEDIUM HAS PRODUCED
FANTASTIC BODY FORMS AND UNUSUAL
LIFESTYLES.

RIGHT MARINE VIRUS
MAGNIFIED 250,000
TIMES. A TEASPOON OF
SEA WATER CAN HOLD
BETWEEN 50,000 AND
50 MILLION VIRUSES.
THESE BITS OF GENETIC
MATERIAL COATED WITH
PROTEIN MOSTLY INFECT
BACTERIA.

OPPOSITE FANCIFUL
LEAFY SEA DRAGONS
LOOK MORE LIKE THE
ALGAE THEY HIDE IN
THAN THEIR FISH RELA-
TIVES. AMONG SEA HORS-
ES AND SEA DRAGONS,
IT IS THE MALE THAT
CARRIES AND INCUBATES
THE EGGS.

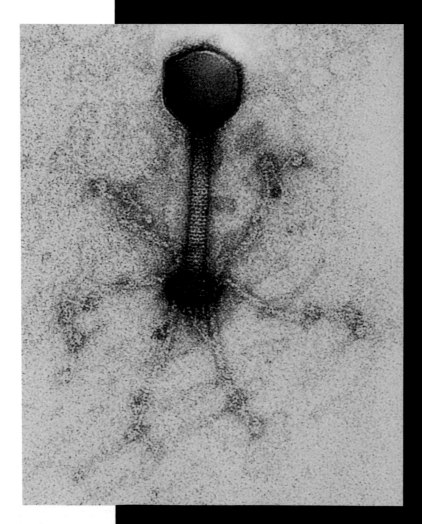

ABOVE LIKE OTHER SEA
CUCUMBERS (RELATIVES
OF SEA STARS AND SEA
URCHINS), THIS SPECIES
CAN SHOOT STICKY POISO-
NOUS TUBULAR THREADS
FROM ITS ANUS. THE
WRITHING TUBULES
ENTANGLE AN ATTACKER,
WHILE THE "VICTIM"
ESCAPES TO REGENERATE
NEW TUBULES.

RIGHT SPONGES ARE
SEDENTARY ANIMALS.
THEY FILTER WATER
THROUGH PORES IN THEIR
BODIES TO "FEED" THEM
DETRITUS AND PLANKTON.

RIGHT PEANUT WORMS LOOK MORE LIKE PEANUTS WHEN THEY CONTRACT THEIR BODIES. ADULTS BURROW IN SAND OR DWELL IN ROCK OR CORAL CREVICES. THE LONG-LIVED LARVAE SWIM IN WARM SURFACE WATERS OF OCEANS THROUGHOUT THE WORLD.

BELOW SEA HARES, UNLIKE THEIR RELATIVES THE SNAILS, DON'T HAVE PROTECTIVE SHELLS. INSTEAD, THEY RELEASE PURPLE INK FOR ESCAPE AND TOXIC WHITE FLUID FOR PROTECTION WHEN DISTURBED.

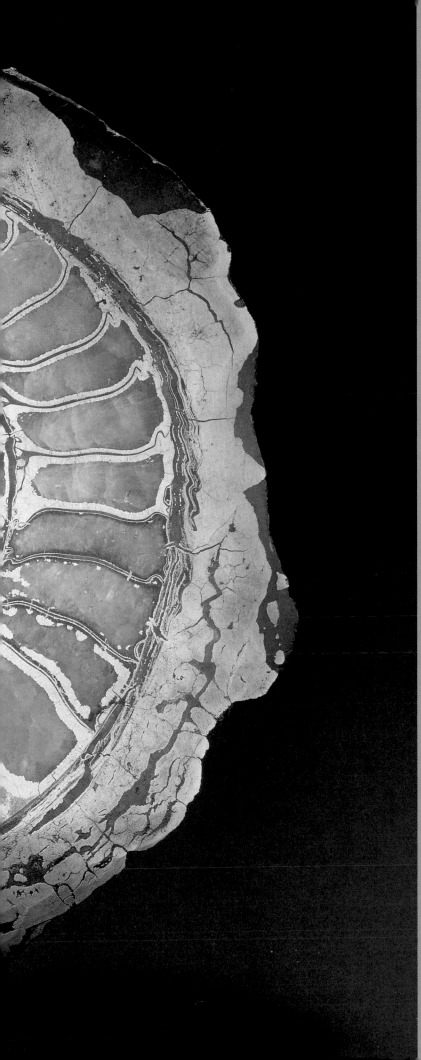

MARINE LIFE FACTS

➤ The oceans contain 99 percent of the living space on the planet.

➤ Hydrothermal vents, fractures in the sea floor that spew sulphur compounds, support the only complex ecosystem known to run on chemicals, rather than energy from the sun. Gigantic tubeworms and mussels thrive in densities of up to 6 pounds per square foot around vents.

➤ The oarfish is the longest bony fish in the world. With its snakelike body sporting a magnificent red fin along its 50-foot length, horse-like face, and blue gills, it accounts for many sea-serpent sightings.

➤ A group of herring is called a seige. A group of jelly fish is called a smack.

➤ Many fish can change sex during the course of their lives. Others, especially rare deep-sea fish, have both male and female sex organs.

The paleontological record is rich in evidence of the antiquity of marine life. This particularly fine fossil ammonite, an extinct relative of the chambered nautilus, from the collection of the National Museum of Natural History, is more than 230 million years old.

➤ Oils from the orange roughy fish — a deep-sea species from New Zealand — are used in making shampoo.

➤ Penguins fly underwater at speeds of up to 25 miles per hour.

➤ Since coral architecture and chemistry are very close to human bone, coral has been used to replace bone grafts in helping human bones to heal quickly and cleanly.

➤ Horseshoe crabs have existed in essentially the same form for the past 135 million years. Their blood provides a valuable test for the toxins that cause septic shock, which previously resulted in half of all hospital-acquired infections and one-fifth of all hospital deaths.

➤ Alginates, derived from the cell walls of brown algae, are used in beer, frozen desserts, pickles, adhesives, boiler compounds, ceramics, explosives, paper, and toys.

➤ The remains of diatoms, algae with hard shells, are used in making pet litter, cosmetics, pool filters, and tooth polish.

➤ One study of a deep-sea community revealed 898 species from over 100 families and a dozen phyla in an area about half the size of a tennis court. Over half of these were new to science.

WHALE STRANDINGS

ONE MEASURE OF HOW LITTLE
WE KNOW ABOUT THE OCEANS
AND THE CREATURES THAT LIVE
IN THEM IS THE FACT THAT
THERE IS NO CONSENSUS ABOUT
HOW MANY SPECIES ARE YET TO
BE DISCOVERED. ESTIMATES
RANGE BETWEEN 500,000 AND
5 MILLION. ANOTHER MEASURE
IS THAT WE HAVE BARELY
BEGUN TO UNDERSTAND THE
BEHAVIOR OF THE SPECIES WITH
WHICH WE ARE FAMILIAR. CON-
SIDER WHALES. WE KNOW VERY
LITTLE ABOUT THEM, HOW
THEY COMMUNICATE, HOW LONG
THEY LIVE, WHAT ILLS THEY
ARE HEIR TO, WHY THEY OCCA-
SIONALLY STRAND THEMSELVES
AND DIE. TO LEARN MORE ABOUT
THE BIOLOGY AND EVOLUTION-
ARY RELATIONSHIPS OF THESE
INTRIGUING ANIMALS, DR. JAMES
MEAD AND HIS COLLEAGUE,
CHARLIE POTTER, OF THE NAT-
IONAL MUSEUM OF NATURAL
HISTORY TRAVEL UP AND DOWN
THE COAST, PERFORMING AUTOP-
SIES AND NECROPSIES, HOPING
TO UNCOVER, IN DEATH, SOME OF
THE SECRETS OF LIFE.

*James Mead found the par-
tial skull, above, of unknown
whale species in Peru in
1976. After biologist Julio
Reyes found more specimens,
including the complete
skull below, he and Mead
described the new species as*
Mesoplodon peruvianus
in 1991.

The whale swam ashore in Nags Head, on
the Outer Banks of North Carolina, shortly before
dusk, even as a group of beachcombers tried to
scare her away by shouting and waving their arms.
Just before she grounded, a dozen people ran
into the water and flushed her back toward open
ocean. But she turned shoreward again and
beached herself solidly in front of a cluster of
cottages.

A hundred yards out, the water was cut by
the vaporous spouts of two other whales, possibly
the female's mate and calf. Bystanders could see
now that the doomed whale was black with a
grayish belly and a dolphinlike nose and about 14
feet long. She was what is called a dense beaked
whale, a little-known creature seldom found
along the mid-Atlantic coast. Her eyes rolled
slowly; her mouth, set in a wry, permanent smile,
remained clamped shut. The tide cut a deep
trough around her as she died.

Thirty-six hours later, James G. Mead, a
scientist from the Smithsonian Institution, stood
over the carcass rhythmically pulling a flensing
knife across a sharpening stone as if preparing
to give the whale a shave. It was low tide on
an August morning. Mead and an assistant had
already attracted a small crowd as they took a
series of measurements — girth, fluke, width, distance from the rostrum to blowhole. Now they
were preparing to cut the whale open.

For the next three hours, Mead indulged
heartily in the messy science that is his great love.

He sliced up the sand-colored blubber to measure
its width, a gauge of the animal's overall health.
He disemboweled the whale with a sickle, expos-
ing the blackish-red meat characteristic of deep
diving species. He piled her intestincs on the sand
like burgundy sausage, cut out her purple stomach
and tied each of it with string so the contents
could be analyzed later. He examined her flaccid
uterus and noted that she had given birth to a
calf only a few months before.

Mead, the 49-year-old curator of marine
mammals at the National Museum of Natural
History, part of the Smithsonian in Washington,
is internationally known for his research on the
anatomy of cetaceans, a group that includes
whales, dolphins and porpoises. Since 1972, he
has built the museum's marine mammal collection
into by far the largest and most complete in the
world. In the process, he has discovered new
species and added significantly to the knowledge
of cetacean natural history.

Anatomy is not a field of science that attracts
popular attention or many lucrative grants. It
offers neither the glamour and intrigue of molecu-
lar biology nor the excitement of studying live
animals in the field. Indeed, many biologists
view it as a discipline more pertinent to Charles
Darwin's time than our own.

And yet, largely because of Mead's work,
anatomy has in the past decade become critical
to the study of these sea animals. By compiling
tremendous stores of data on every aspect of

marine mammals — their growth rates, reproductive history, skeletal dimensions and the toxins contained in their flesh — he has given scientists a detailed look at species that often can be observed only fleetingly in the wild, where they swim too quickly and dive too deep to be closely studied. Although some species have been photographed extensively and even fitted with radio transmitters and tracked, many researchers believe the process may affect their natural behavior.

Mead has also left his mark on the field through his prowess as an investigator. He has amassed an impressive body of research, especially on beaked whales, which have long snouts like dolphins and are among the most mysterious creatures on earth. Scientists believe beaked whales dive deeply, perhaps to three or four miles, and stay submerged for hours at a time. In 1991, Mead announced the discovery of a previously unknown species of beaked whale, the pygmy beaked whale, *Mesoplodon peruvianus*. Although he found the first pygmy carcass in 1976, it took him 15 years to obtain enough evidence to describe the species in detail. . . .

When Mead accepted the job at the museum, he assumed he would be able to collect whales from the Canadian fishing industry and perhaps from other countries that still allowed commercial whaling. But after the 1972 season, Canada abruptly outlawed whaling. Then in December, Congress passed the Marine Mammal Protection Act, one of the most restrictive environmental laws in the world. "It suddenly became nearly impossible to collect marine animals, no matter how lofty our intentions or how great our scientific need," Mead says. "Marine mammals are so heavily protected that it's analogous to doing research on humans — except you can't ask a dolphin or a whale why it's sick or where it hurts."

The museum collection stood at about 2,000 specimens. To expand it, Mead decided he would have to find stranded whales. If an animal beached "within a day's drive of Washington,"

Mead and Charlie Potter set out immediately to retrieve it. "We defined 'day's drive' as 24 hours of straight driving," Mead says. "So we covered the East Coast from Cape Cod to Charleston. We were on the road constantly." Whales were hauled onto a flatbed truck and driven to Washington.

Occasionally Mead flew to a stranding on a commercial airline, which had certain drawbacks, particularly on the return trip. Once in the mid-1970's, he flew to Cape Fear in North Carolina to dissect a stranded Gulf Stream beaked whale. He packed the whale parts into plastic garbage bags and checked them as excess baggage. "You don't want to advertise the fact that your traveling companion is a dead whale," he says. "We got back to National Airport and my baggage began to come out and it smelled a little funny. And it was leaking something very nasty looking." Oil within the bones of the whale had penetrated the plastic bags. "A guy pulled up on a baggage cart with more leaking bags and asked what was in them. Before I could answer, he said, "Looks like hydraulic fluid. Looks like you got a problem, buddy." I just nodded. If he wanted to think they were machine parts leaking hydraulic fluid, that was beautiful. . . . "

Much of Mead's anatomical work has proved useful to fisheries managers who are struggling to balance marine conservation with the needs of commercial fisherman.

Last summer, Mead and Potter discovered a subtle difference in the nasal passages of Atlantic bottlenose dolphins that live along the coast and those that live far off-shore. Scientists had long suspected that the dolphin population consists of two separate stocks, but they had no way to prove it or to study the behavior of the two groups until Mead and Potter's discovery.

"If you have one big population of bottlenose dolphins in the mid-Atlantic, it's much easier to say, O.K., we'll allow so much incidental catch," Mead says. "But if you have two distinct stocks that don't interbreed, as we suspect, you can't allow as many dolphins to be killed by fishing gear or else the whole population's going to suffer."

Mead is also at work on a study of dolphin reproduction. By examining scars on the uterus and mammary glands, he has found that females often do not give birth until the age of 10, though they first ovulate around age 6. This means their reproduction rate is much lower than previously thought.

The reproductive study is especially pertinent in light of the massive die-off of bottlenose dolphins in 1987 and early 1988, when 1,200 stranded between New Jersey and Florida. "That's probably only a tenth of the number that actually died," Mead says. "The others just didn't wash up on the beach." Some scientists now think that half the mid-Atlantic population may have died during that period. No one knows what caused the deaths, although some researchers believe environmental toxins or a poisonous plankton may

have caused the animals' immune systems to fail. The question of how quickly the dolphin population can regain its numbers will be a key consideration as national officials debate whether to place restrictions on fishing activities harmful to dolphins, like trawling.

"People traditionally think of museum work as being limited to collecting bone samples," says Aleta Hohn at the National Marine Fisheries Service in Silver Spring, Md., who collaborated with Mead on the study. "Jim and Charlie Potter have created a tremendous repository of information by looking at all parts of the animals they collect — the stomachs, the gonads, the mammaries, the heads. Their work is really exceptional."

— JAN DeBLIEU, "The Messy Science of Cetology," *The New York Times Magazine*, February 21, 1993

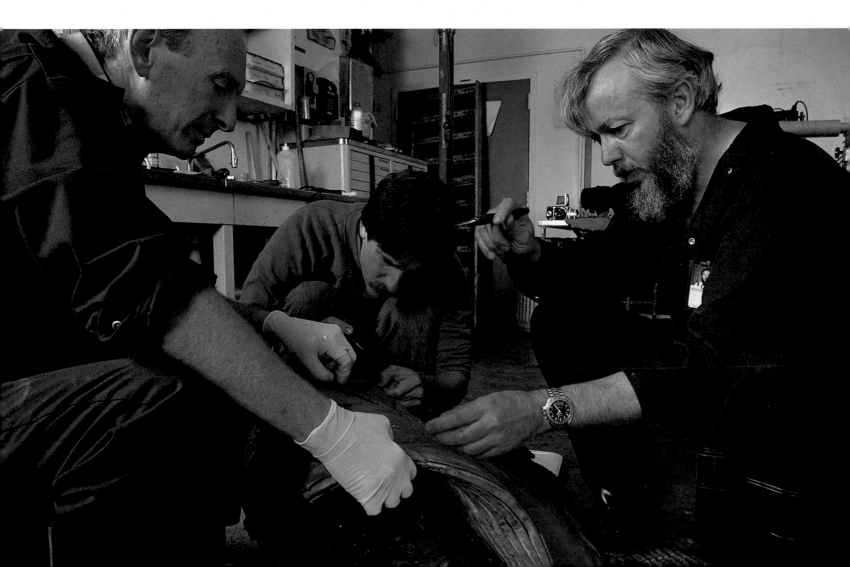

RETURN TO THE SEA

LIFE ON EARTH BEGAN IN THE SEAS OVER
THREE BILLION YEARS AGO, AND STAYED
THERE UNTIL ABOUT FOUR HUNDRED MILLION
YEARS AGO, WHEN THE ANCESTORS OF REP-
TILES CRAWLED OUT ONTO LAND. SINCE THEN,
OVER EVOLUTIONARY TIME, SEVERAL GROUPS
OF PREDOMINANTLY LAND-DWELLERS MOVED
BACK TO THE SEA TO LIVE.

THE AUSTRALIAN SEA
LION IS A SPECIES
ESTIMATED AT 3,100
TO 6,900 INDIVIDUALS,
FOUND MOSTLY ON
THE SOUTHERN COAST-
LINE OF THE CONTINENT.
MANY BEAR SCARS
FROM SHARK ATTACKS.

SIBLE FOR LUSH KELP
BEDS, BECAUSE THEY
EAT SEA URCHINS AND
ABALONE THAT GRAZE ON
KELP HOLDFASTS. WHEN
NINETEENTH-CENTURY
HUNTING ALMOST ELIMI-
NATED CALIFORNIA
SEA OTTER POPULATIONS,
URCHIN AND ABALONE
POPULATIONS INCREASED,
DESTROYING SOME GIANT
KELP FORESTS.

SEA TURTLES LIVE AT
SEA BUT THEY RETURN
TO LAND TO LAY EGGS.
THEY HAVE AN UNCANNY
SENSE OF DIRECTION,
AND SOME MIGRATE OVER
A THOUSAND MILES TO
RETURN TO NESTING
BEACHES. SEA TURTLES,
LIKE THIS GREEN TURTLE,
ARE HUNTED FOR
THEIR MEAT AND THEIR
SHELLS. MANY DROWN
IN FISH NETS.

THE BLUE WHALE IS THE
LARGEST KNOWN ANIMAL
EVER TO HAVE LIVED ON
SEA OR LAND. INDIVI-
DUALS CAN REACH OVER
110 FEET AND WEIGH
NEARLY 200 TONS —
MORE THAN TWENTY
ADULT ELEPHANTS. THE
BLUE WHALE'S BLOOD
VESSELS ARE SO BROAD
THAT A FULL-GROWN
TROUT COULD SWIM
THROUGH THEM, AND THE
VESSELS SERVE A HEART
THE SIZE OF A SMALL
CAR. DESPITE ITS SIZE,
THE BLUE WHALE FEEDS
ON TINY PHYTOPLANKTON.
AN ADULT SUCKS IN 45
TONS OF WATER IN A
GULP AND FILTERS OUT
THREE TO FOUR TONS OF
KRILL AND SMALL FISH
DAILY.

MAPPING THE SEABED

THE EXCLUSIVE ECONOMIC ZONE OF THE UNITED STATES IS A BELT OF WATER THAT SURROUNDS THE LAND MASS OF THE NATION AND EXTENDS TWO HUNDRED MILES OFFSHORE. ALTHOUGH WORLDWIDE ONLY ABOUT 5 PERCENT OF THE OCEAN BOTTOM HAS BEEN PLOTTED IN DETAIL, FOR MORE THAN A DECADE THE UNITED STATES HAS BEEN CONDUCTING A PROJECT TO MAP ITS ENTIRE EEZ, THE LARGEST ON THE GLOBE. THE BENEFITS OF PRECISE MAPPING, WHILE AS YET UNCERTAIN, ARE POTENTIALLY ENORMOUS. AMONG THE MINERAL RESOURCES KNOWN TO EXIST ON THE SEA BED IN VAST QUANTITIES ARE MANGANESE, NICKEL, COPPER, AND COBALT. THE COARSE-SCALE MAPPING IS DONE BY SATELLITES, BUT THE FINE WORK IS PERFORMED DAY AFTER DAY, MILE AFTER MILE, FROM A SHIP MANNED BY PATIENT EXPERTS — AND BY AN UNCOMPLAINING WORKER KNOWN AS GLORIA.

More formally called the Geological Long Range Inclined Asdic, GLORIA — seen here being hauled out of the water — is a British Geological Survey sonar tool for mapping the sea floor. Sounds emitted from GLORIA reveal bottom topography when they bounce off the bottom and return.

This cruise in the *Farnella* is one of a series at the end of a great project on the part of the US to map the 200-mile EEZ around all 19,924 km of its coastline. The EEZ of the continental US has already been mapped, including Alaska and the Aleutian Islands. Now *Farnella* is working away at the Hawaiian chain. As the scientists aboard keep telling me, only the GLORIA system could have covered such an area, and even that has taken ten years. . . .

Time was spent checking the various instruments which were soon to be lowered into the sea and towed behind us. First of all GLORIA itself: a large yellow torpedo lined inside with banks of transponders precisely angled to give the correct fan-shaped pulse. It sits in a hydraulic cradle directly over the ship's stern. From time to time technicians climb up to tighten a nut and pat it protectively. Even without its cable it is worth nearly half a million pounds. There is spare cable but only one GLORIA aboard. "We don't even like to think about that," is the response to the obvious "What if?" . . .

Over the port stern will go a magnetometer to measure magnetic variability in the Earth's crust, and down in the lab is a gravimeter to record differences in its gravitational field. This machine looks, and is, expensive. It is suspended in a cradle mounted in computer-controlled gimbals, dipping and tilting so it appears to be the one thing in the lab which is constantly in motion, whereas it is really the only thing aboard remaining utterly still while the ship gyrates about it. At supper the conversation turns to where might be the best place on Earth for setting high-jump records, a particular spot with significantly weaker gravity. All the best ones seem to be covered by a couple of miles of water. In response to a remark of mine which betrays real ignorance about gravity, Roger says kindly:

"I suppose one always imagines the surface of the oceans as basically flat. Ignoring waves and local storms, of course — they're just 'noise.' But apart from its being curved to fit the surface of the globe, one thinks of the sea as having to be flat because at school we're told water always finds its own level so as to be perfectly horizontal. On a small scale that's pretty much true, though when I was about ten I remember being surprised when someone pointed out that all rivers are tilted, and if you row upstream you're also rowing uphill as well as against the current. Anyway, since gravity varies from place to place it acts variably on the sea, too. When you start using instruments like the ones aboard this ship you really appreciate how the ocean surface actually dips and bulges all over the place. It shows up best from space."

He explains that by having enough satellites in orbit making passes over the same area, day after day for months on end, it was possible to build up a mean reading for the height of the sea's surface at that spot. It took a long time because there was a good deal of "noise" to be discounted: wind heaping, sudden areas of low atmospheric

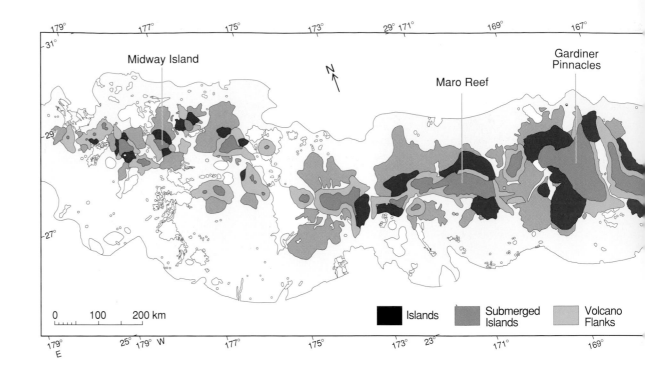

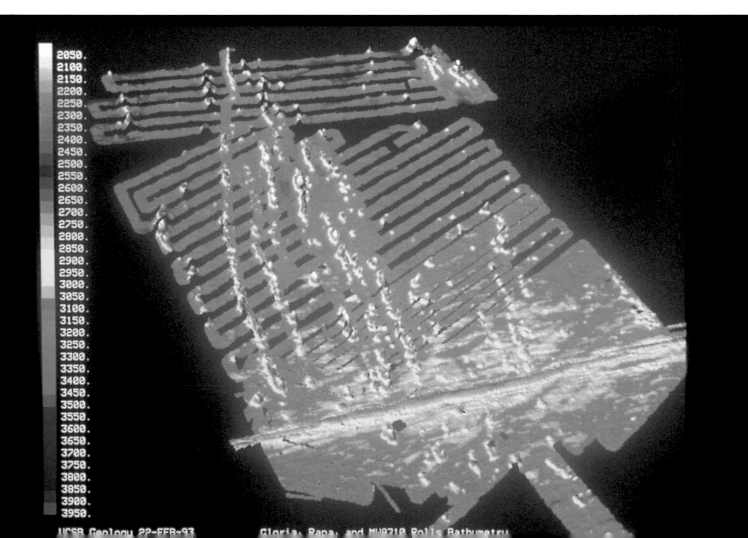

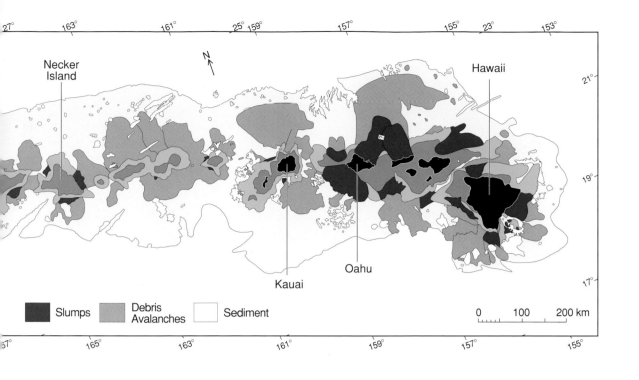

Necker Island

Hawaii

Kauai

Oahu

| Slumps | Debris Avalanches | Sediment |

0   100   200 km

BELOW *A computer-generated image showing the topography of three sea-mounts in the East Pacific Rise. With data from* GLORIA, *oceanographers can generate relatively accurate topographical maps.*

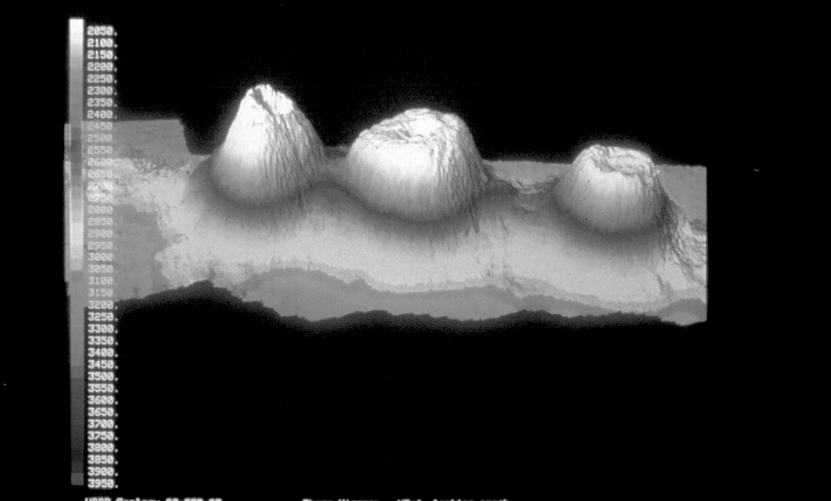

2050.
2100.
2150.
2200.
2250.
2300.
2350.
2400.
2450.
2500.
2550.
2600.
2650.
2700.
2750.
2800.
2850.
2900.
2950.
3000.
3050.
3100.
3150.
3200.
3250.
3300.
3350.
3400.
3450.
3500.
3550.
3600.
3650.
3700.
3750.
3800.
3850.
3900.
3950.

UCSB Geology 22-FEB-93   Three Wisemen - VE=4, looking north pressure which could suck the sea upward as if beneath a diaphragm, even very low-frequency waves with swells so long they might take half a day to pass. But if the satellites went on measuring the same spot for long enough, such fluctuations would even out and a geodetic point be established: a mean distance to the sea's surface as measured from the center of the Earth. By building up enough geodetic points it soon became clear that the oceans were anything but flat.

"What's more, if you match this up with the underlying features on the seabed, you'll find that the surface of the sea broadly mimics the topography underneath. And the reason for *that* is fluctuations in gravity, which depends on the density of the crustal material."

It is a pretty notion, that the sea follows the Earth's crust like a quilt laid over a lumpy mattress. It is also odd to think that to some extent the depths of the oceans can be read from space. . . .

As is all too clear to anybody swimming in circles looking for a lost boat in the middle of the ocean, one has no position in water. When mapping the seabed from a moving ship, therefore, accurate navigation is of crucial importance. Without the ship's position being known from one second to the next the most beautiful chart of peaks, ravines and plateaus would be useless. The only thing known would be that they were down there somewhere. Establishing the ship's course along lines as straight as possible (always allowing for the Earth's curvature) requires much work, not least because the swaths GLORIA maps must lie next to each other without gaps or wasteful overlapping. On the chart table in the lab is the dot of Johnston Island, a pencil circle whose diameter represents 400 nautical miles inscribed about it. High up in its top left-hand quarter a chord shows the first leg we have just started. Next to it is written the estimated time at which we should come about for the return pass, each leg getting longer as we eat downward into the circle. If all goes well,

by the end of a fortnight we should have hatched off about a quarter of the total area.

While the lab computers flicker with the instruments' returning signals, various repeater gauges give the ship's speed through the water, its speed over the ground, the wind speed and any consequent degree of yaw. If to remain on a straight course against a quartering wind and current the *Farnella* needs to sail crabwise, GLORIA's angle will also be fractionally oblique to its correct path. The result is that its signals will no longer be exactly at right angles to this course and the map will be distorted. Information on all these factors is fed into the computers, which correct for them. In order to determine the ship's position at any moment the *Farnella* uses GPS, or Global Positioning System. This depends on satellites and eventually, provided there are still spare slots in an already overcrowded geostationary orbit, the system will cover the Earth and in theory allow a person anywhere on the planet's surface to determine his position to within a few meters. This would not be of the slightest use to a lost swimmer looking for his boat.

Bored with the sight of bright red digital figures flickering their decimal points on display panels I wander off in search of sound. . . .

Very occasionally from a chance position down in the hull, at some freak acoustical window, it is possible to hear GLORIA's peculiar yodel. The instrument emits a correlation signal; instead of a single bleep its pulses take the shape of a whistle which swoops up and down. This is so the echo will be unmistakable, the electronic ears listening for its return being tuned to exclude all other signals. Even so, knowing how to read the GLORIA trace as it emerges from the plotter is a matter of much experience. . . .

The real distinction between this kind of oceanography and all that went before is not merely that the technology has changed, and with it the techniques for analyzing data. It is that the scientists themselves are using different senses.

Nobody is actually listening to these signals returning from unexplored regions laden with information. The lab is filled, not with the hollow pinging familiar from submarine war film soundtracks, but with the click and whir of plotters and jocular bouts of repartee. No one now wears headphones and a rapt, faraway look, attentive in ambient hush. For all that modern oceanography relies so much on acoustic techniques, it is machines which do the listening. . . .

The next day it is discovered that existing charts of this area, which like most maps (other than those of pirate treasure) look completely authoritative, are quite wrong. Whole features are either absent or misplaced. We examine the printouts. Entire mountains flicker in and out of existence somewhere down there in the cold darkness. The bathymetry is all haywire. GLORIA knows best. There remains the experience of being present when a portion of the Earth's surface is discovered. This is a rare sensation for the layman in the late twentieth century, and the diminishing opportunities for experiencing it must belong almost exclusively to potholers and cavers, apart from oceanographers themselves. . . .

The days pass, the ship goes back and forth along its lines, the shaded portion of the circle grows. No disasters, only the single hitch of the weather. The wind is stiff, the waves high, as if we were perpetually on the edge of a storm system 500 miles away. The seas hit us at an angle, nudging the bows off course. There are worries about possible damage to GLORIA's cable, which thrums like a steel bar over the stern, slackening in troughs and then tautening with a snap. The scientists confer with the captain and agree to knock a knot off the speed. This means computer work so the scans are not distorted. A new seamount is discovered whose foothills were first spotted on the previous leg while traveling in the opposite direction.

"I don't know what the hell that is," says Mike, pointing at a gravelly-looking portion of a gravelly-looking picture. Doctors poring over an X-ray.

"I reckon it's a nodule field," says Roger. "I've seen something just like it elsewhere. Those are really sheer cliffs [rock faces of 3,000 feet which the world's mountaineering community would race to get their pitons into were they on dry land], and there's less sedimentation here so we're in the lee of the local current system. That means *this* stuff" — a chewed ballpoint taps the picture — "is boulders. Detritus that's fallen over the edge or sheared from the face. Doesn't look like any new activity here, so these'll be nodules, half buried by the debris and extending out to — wow, it may be off this leg too, so it's a big field. Anyway, bet you anything that's what it is. I feel it in my bones."

"And can Roger's famous bones be wrong?" murmurs Mike.

— JAMES HAMILTON-PATERSON,
The Great Deep: The Sea and its Thresholds, 1992

OCEANS IN PERIL

MINAMATA

In the 1950s, residents of Minamata, Japan, began to suffer from strange and devastating ailments, including nervous-system disorders and birth defects. In all, more than 2,200 people became ill, and at least 730 died, from what became known as Minamata disease. They had all eaten seafood contaminated by industrial wastes containing methyl mercury, which were traced to a plant belonging to the Chisso company. Mercury poisoning is among the most severe, but by no means the only, result of marine pollution. Untreated sewage pouring into coastal waters can carry hepatitis, cholera, typhoid, and a wide range of other diseases and infections. Clean-ups are often extremely difficult, if not impossible, and slowly degrading contaminants can leach from sediments for hundreds, even thousands, of years.

A woman holding her daughter, who is suffering from Minamata disease, early 1970s.

My name is Asae Fukuda, and I am a Minamata disease patient.

I am appearing in court on behalf of my daughter Itsuko, who died at the age of 11. My home is near a fishing port called *Hakariishii Port,* located ten odd kilometers north of the Chisso Plant.

I became pregnant in 1967. As this would be the birth of our first child, family and relatives were all very happy.

Thinking that it would be for the good of the child within me, I ate fresh fish every day. My happiness, however, was short lived.

Even seven months after birth, Itsuko could not support the weight of her own head, and could not see. Many times a day she suffered convulsions, contorted her infant features, and drooled, making it unbearable to watch.

When she became one year old, I was told at the Kumamoto University Hospital that Minamata disease was suspected.

I did not think it could be so, and did not want to believe it, but the doctor at the University Hospital asked me if I was living near the sea, and if I ate a lot of fish. Everything he asked was so, and I could only accept that he was right. It was then that I began taking Itsuko to the hospital regularly.

Every time I heard of a good doctor, I took Itsuko to see him. Desperate to try anything, I also visited temples and shrines to pray.

Three or four times a month I took Itsuko to Kumamoto University Hospital for examinations, a trip that took two hours each way. On the days after examinations Itsuko was especially tired, and would cry all night. At such times I would carry Itsuko on my back to the nearby port to humor her. I cried with my child, who would not understand if I talked to her, as I thought of how great a relief it would be for us to throw ourselves in the sea and die.

When Itsuko was three, I wanted to dress her in fine clothing for the *Shichigosan* children's day, though she could not stand on her own. Even if I put the clothing on Itsuko, who was lying down, she would only be like an unmoving doll. It was so sad that I cried as I dressed her. I then put makeup on her as well. As I carried her on my back to the shrine I said, "Itsuko, when you're able to stand let's go once more to the shrine for *Shichigosan.*"

At mealtime I gave her liquid food spoonful by spoonful, being very careful, as if praying, that she would not choke on it.

I was assailed by the thought that I made Itsuko this way because of the fish I ate durin pregnancy. So I believed that I must care for her whatever should happen, and I nursed the child in her illness with all my heart.

However, during the winter when she became 11 years old, in February when it was still cold, Itsuko died while suffering respiratory difficulty.

There was more unhappiness. A child con-

Waste emptying into Minamata Bay, early 1970s. The greatest mercury discharges occurred during the 1950s, when production of acetylaldehyde and vinyl chloride was at its height, and the population of Minamata was 50,000 people.

ceived in the fifth year was stillborn after five months. There are many women around me in Minamata who have experienced stillbirths and miscarriages. Chisso's mercury robbed us mothers of our greatest joy.

But the governor of Kumamoto denied that Itsuko had Minamata disease. I could not acccept that, and I am thus a member of the plaintiffs in the Third Minamata Disease Lawsuit.

In a decision of last March 30th, the Court recognized that Itsuko had fetal Minamata disease. Furthermore, the Court said that the National Government and Kumamoto Prefecture shoulder this grave responsibility with Chisso.

How long I had waited for the day when I would hear this!

Just after the decision had been handed down, I went to Tokyo to negotiate with the Chisso main office and the Environment Agency.

Anger welled up within me during these negotiations toward the attitude of Chisso and the government, who would not discharge their responsibility in accordance with the Court decision. Why should those of us who have suffered this harm be subjected to such mortification? To Itsuko, who had suffered 11 long years, I wanted them to say, "What a shame we caused such a thing. We're sorry." Since that time I have negotiated with them many times, but still they take no responsibility.

However, I think our struggle has made considerable progress, for a small change is visible in the attitudes of Chisso and the government.

Last October I went to the United Nations headquarters in New York with many supporters. The purpose was to ask the UN Human Rights Commission for a redress of human rights for Minamata disease victims.

Since that was of course the first time I had gone abroad, I felt anxiety that I would not be able to convey my thoughts to people with different skin color, languages, customs, and ways of thinking. However, the people in the responsible departments and bureaus listened to my appeal with tears in their eyes. The fears I had felt until that time were alleviated, and I believed without a doubt that we had communicated our common sentiments as human beings. This May I received a notice from the UN that my appeal had been officially accepted. I think, however, that the struggle has only begun.

I have been able to persevere this long because of my 11 years of suffering.

I cannot forgive the National Government, Kumamoto Prefecture, and Chisso, who robbed Itsuko of her life, and me of my health.

Whenever I think I am about to give up, I see Itsuko's face, continuing the struggle to live even as she fights to breathe, and I can hear her voice saying, "Mommy, don't give up the fight."

I shall continue this fight as long as I live.

Thank you for your attention.

— Statement by Ms. Asae Fukuda to the International Forum on Minamata Disease held in Kumamoto City, Japan, November 7–8, 1988

DUMPING AT SEA

FOR THOUSANDS OF YEARS HUMANS HAVE
VIEWED OCEANS AS VAST DUMPS FOR DOMES-
TIC, MUNICIPAL, AND INDUSTRIAL GARBAGE,
INCLUDING SEDIMENTS DREDGED FROM HAR-
BORS AND WATERWAYS, SEWAGE SLUDGE,
TOXIC INDUSTRIAL BY-PRODUCTS, AND EVEN
LOW-LEVEL RADIOACTIVE WASTE. SOME OF
THIS MATERIAL MAY NEVER BE DILUTED INTO
A UNIFORM BACKGROUND CONCENTRATION,
AND OCEAN PROCESSES MAY EVEN CONCEN-
TRATE SOME COMPOUNDS. HOWEVER, LAND-
BASED ALTERNATIVES FOR WASTE DISPOSAL
ALSO POSE PROBLEMS.

M/V *ANSON* DUMPING
JAROSITE AT SEA, APRIL,
1990. AUSTRALIA HAS
PERMITTED DUMPING OF
JAROSITE, A BY-PRODUCT
OF ZINC SMELTING,
SINCE 1973.

THREATS TO THE OCEANS

➤ Oil spills account for only about five percent of the oil entering the oceans. The Coast Guard estimates that for United States waters sewage treatment plants discharge twice as much oil each year as tanker spills.

➤ Each year in the United States, we dredge 400 million cubic yards of sediment from channels and harbors to keep them navigable — the equivalent of a four-lane highway 20 feet deep from New York to Los Angeles. About 2–15 percent of this material contains hazardous chemicals.

➤ It is estimated that medical waste that washed up onto Long Island and New Jersey beaches in the summer of 1988 cost as much as $3 billion in lost revenue from tourism and recreation.

Tourists enjoy the Mediterranean near Cannes, France, 1991. By the mid-1970s the heavily populated coastline of the Mediterranean Sea was a mess: tar balls washed up on beaches and raw sewage contaminated coastal waters. In 1975, eighteen countries bordering the sea adopted the United Nations-sponsored Mediterranean Action Plan to reduce land-based sources of pollution throughout the region.

➤ The most frequently found item in beach cleanups is pieces of plastic. The next four items are plastic foam, plastic utensils, pieces of glass, and cigarette butts.

➤ Lost or discarded fishing nets keep on fishing. Called "ghost nets," this gear entangles fish, marine mammals, and sea birds, preventing them from feeding or causing them to drown. On average, as many as 20,000 northern fur seals may die each year from becoming entangled in netting.

➤ Air pollution is responsible for almost one-third of the toxic contaminants and nutrients that enter coastal areas and oceans.

➤ When nitrogen and phosphorus from sources such as fertilizer, sewage, and detergents enter coastal waters, oxygen depletion occurs. One gram of nitrogen can make enough organic material to require 15 grams of oxygen to decompose. A single gram of phosphorus will deplete one hundred grams of oxygen.

➤ The Mississippi River drains more than 40 percent of the continental United States, carrying excess nutrients into the Gulf of Mexico. Decay of the resulting algal blooms consumes oxygen, kills shellfish, and displaces fish in a 4,000 square mile bottom area off the coast of Louisiana and Texas, called the "dead zone."

➤ In 1993, United States beaches were temporarily closed or swimmers advised not to get in the water over 2,400 times because of sewage contamination. The problem is even worse than the numbers indicate: there are no federal requirements for notifying the public when water-quality standards are violated, and some coastal states don't monitor water at beaches.

➤ The zebra mussel is the most famous unwanted ship stowaway, but the animals and plants being transported to new areas through ship ballast water pose a problem around the world. Poisonous algae, cholera, and countless plants and animals have invaded harbor waters and disrupted ecological balance.

➤ There are 109 countries with coral reefs. Reefs in 90 of them are being damaged by cruise ship anchors and sewage, by tourists breaking off chunks of coral and sea fans, and by commercial harvesting for sale to tourists.

➤ One study of a cruise ship anchor dropped in a coral reef for one day found an area about half the size of a football field completely destroyed, and half again as much covered by rubble that died later. It was estimated that coral recovery would take fifty years.

➤ Egypt's High Aswan Dam, built in the 1960s to provide electricity and irrigation water, diverts up to 95 percent of the Nile River's normal flow. It has since trapped more than one million tons of nutrient rich silt and caused a sharp decline in Mediterranean sardine and shrimp fisheries.

➤ Commercial marine fisheries in the United States discard up to 20 billion pounds of non-target fish each year — twice the catch of desired commercial and recreational fishing combined.

➤ Almost half of all construction in the United States during the 1970s and 1980s took place in coastal areas.

➤ Within thirty years a billion more people will be living along the coasts than are alive today.

➤ By the year 2000, thirteen of the world's largest cities will lie on or near coasts.

➤ With only 4.6 percent of the world population, Americans use about one-third of the world's processed mineral resources, and about one-fourth of the world's non-renewable energy sources, like oil and coal.

BERNIE FOWLER DAY

A gricultural pollution by most measures is at least not getting worse, but the same can't be said for population growth. I'm thinking of this as I gaze up the Potomac River on a hot June afternoon. Plenty of time to gaze, because I'm caught on the Woodrow Wilson Bridge, just below Washington, D.C., barbecuing in some of the world's worst traffic congestion, inhaling fumes that are an acrid affront to the nation's Clean Air Act.

THE CHESAPEAKE BAY IS ONE OF THE MOST BELOVED AND CELEBRATED ESTUARIES IN NORTH AMERICA. FED BY 150 TRIBUTARY RIVERS AND STREAMS, ITS WATERSHED EXTENDS OVER 64,000 SQUARE MILES, AND IT HAS LONG BEEN ONE OF THE WORLD'S MOST PRODUCTIVE BODIES OF WATER, SHELTERING FISH AND SHELLFISH IN GREAT ABUNDANCE. FOR THE PAST FEW DECADES, HOWEVER, THE CHESAPEAKE HAS BEEN TROUBLED. WATER QUALITY AND WILDLIFE HAVE DECLINED AS HUMAN POPULATION AND POLLUTION HAVE INCREASED. A MAJOR CLEAN-UP EFFORT IS UNDERWAY, BUT IT IS COMPLICATED BY THE VERY SIZE OF THE BAY, WHICH CROSSES POLITICAL BOUNDARIES AND ADMINISTRATIVE JURISDICTIONS. AMONG THE LEADERS OF THE CRUSADE IS A CHARISMATIC FIGURE NAMED BERNIE FOWLER, WHOSE GOAL IS TO LOOK DOWN INTO THE CHESAPEAKE'S WATERS AND SEE A SIGHT HE HASN'T SEEN SINCE CHILDHOOD: HIS TOES.

Nanticoke River, Chesapeake Bay, 1992. Excessive amounts of nitrogen and phosphorus from pollution in the Nanticoke River resulted in overpopulation of plants and algae. When the plants and algae died, decomposing bacteria removed too much oxygen from the water, killing the fish.

The river below us, the bay's second largest tributary, once supported oyster beds downstream so rich that dozens of Marylanders and Virginians have died in shoot-outs over the bounty in the past century. Sturgeon, measuring as long as 14 feet, once roiled the river here, where a caviar business operated until the mid-1920s.

While still not back to its historical levels of production, the Potomac is cleaner today than it was in the 1960s, when President Lyndon B. Johnson called the nation's river "disgraceful." Johnson's warning led to massive spending on sewage treatment, but as population grows in the Washington region, the pressures on the bay will continue mounting.

Bernie Fowler saw the crush of humans — and trouble — coming more than 20 years ago, when he was a new president of the Board of County Commissioners in Calvert County, Maryland, where the Patuxent River flows down to its junction with the Chesapeake. Bernie mystified many neighbors in his sleepy, rural

county back then by saying that growth was the region's most pressing problem.

"All you had to do was look at what was happening around Washington, D.C., and look at the bridges and highways being planned," Bernie told me recently. "We had waterfront in Calvert County. We had low taxes. We had a beautiful countryside, and everything we were doing was making it more convenient for people to come here."

Since 1950, population in the Patuxent drainage basin has more than tripled, while forests and fields have been given over to development, removing some of the natural buffers that keep pollutants out of the bay.

"You wonder how we let it go," Bernie mused. "You know, you hear people say the water seems cloudy, and I wonder what's happened to all the hardheads [croakers, a once plentiful bay fish]. The fishing fell off. The bay grasses disappeared. You notice all these things, but it comes so slow. I guess it's like setting in a room and the oxygen being consumed; you don't notice it until most of it's gone."

To stem the tide, Bernie and others joined in a lawsuit against Maryland and the EPA in 1978. The idea was to limit sewage pollutants from urban areas upstream. Bernie's side won, and that is why nowadays, every second Sunday in June, people from around the state of Maryland wade into the Patuxent on Bernie Fowler Day. They are hoping to see their toes in the river, just as Bernie

Bernie Fowler (center) and friends testing the water of the Patuxent River on the second Sunday in June, 1993. Fowler vows to continue the annual ritual until the water is so clear he can wade up to his chest and see his toes.

did as a young man in the 1950s. He recalls how he would wade out into the river to hunt for soft crabs. The water was so clear that, even chest deep, he could spy the crustaceans hiding in pale green grasses that carpeted the shallows. Thus the unofficial goal: When Bernie Fowler can once again wade to his chest and see his toes, we will know the river is back.

With Bernie and a crowd of others, I link arms and wade out from the beach at Broomes Island, where flags flap and big fluffy white clouds race before a smart southwest breeze, wafting the smell of a fried-chicken picnic awaiting us on shore.

Shin deep now, I can still see my white tennis shoes clearly (it helps to have size 15 feet). Bits of aquatic grass float by. Better than it used to be, a local tells me. Knee deep now. I don't think we're going to make it, although there is progress. Phosphorus from sewage discharge is down 75 percent since 1981; tens of millions of dollars have been allocated to begin removing nitrogen. Thigh deep and we're done — still looking for our toes, still hopeful.

I've heard Bernie speak many times, but his talk this year is more urgent, perhaps because of the quadruple-bypass surgery he underwent following last year's wade-in.

"When people like me grow old and die off, we risk leaving a whole generation that has no idea what this river really was. No memory banks in those computers at EPA can recall the ten barrels of crabs one person used to catch out there, and all the hardheads, and the thrill of the oyster fleet coming in at sunset, the shuckers in the oyster house all singing harmony while they worked. If we can't make some headway soon," Bernie added wistfully, "these children will never, never have the hope and the dream of bringing the water back, because they just won't have any idea how enriching it used to be."

Even if Bernie finds his toes eventually, in many regions of the river the look of the landscape, the mix of villages and tobacco barns, forest and field and rolling vistas that made southern Maryland unique — much of that is beyond recall.

Increasingly I look at the Patuxent and think of other bay rivers — Virginia's James and Rappahannock, where forested shorelines still predominate for miles at a stretch and you can see more eagles in spots than most places outside of Alaska; the Delmarva Peninsula's Pocomoke and Nanticoke, where you can canoe all day with scarcely a sight of human commerce. These rivers still have time to do it right, to plan for growth rather than react to it, to maintain the great green filters — forest and wetlands — that help cleanse runoff before it reaches the bay. Those riverside communities should send someone to Bernie Fowler Day.

— TOM HORTON, "Chesapeake Bay — Hanging in the Balance," *National Geographic,* 1993

The Chesapeake Bay watershed is a mosaic of forest, agricultural land, and suburban and urban development. It has been one of the world's most productive natural areas, but is increasingly threatened by pollution and over-development.

COASTAL DEVELOPMENT

AMERICA'S COASTAL POPULATION WILL
GROW TO MORE THAN 127 MILLION PEOPLE
BY 2010 — AN INCREASE OF MORE THAN
60 PERCENT IN ONLY FIFTY YEARS. COASTAL
DEVELOPMENT HARMS NATURAL COASTAL
ECOSYSTEMS AND PRODUCES POLLUTION.

RIGHT HOUSING
CROWDS NEW JERSEY'S
BEACH ISLAND STATE
PARK.

OPPOSITE ON MUCH
OF THE ATLANTIC AND
GULF COASTS OF THE
UNITED STATES, BARRIER
BEACHES SHIELD THE
MAINLAND FROM STORMS,
SEA LEVEL CHANGES,
AND EROSION. THIS
PHOTOGRAPH SHOWS
THE IMPACT OF A 1992
NOR'EASTER ON A
STRETCH OF BARRIER
BEACH AT WESTHAMPTON
BEACH, NEW YORK:
HOUSES THAT WERE ONCE
UP ON THE BEACH ARE
NOW STANDING IN THE
WATER, AND A NEW
INLET HAS BEEN OPENED
UP TO THE BAY.

TIM BENTON'S HARVEST FROM DUCIE ATOLL

HE COMPILED THIS LIST OF
BEACH DEBRIS AFTER SURVEYING
THE SEAWARD EDGE OF ACADIA,
THE LARGEST ISLET IN THE
ATOLL (ON THE RIGHT IN THE
AERIAL PHOTOGRAPH ABOVE).

BUOYS: LARGE (USUALLY 30CM DIAMETER)	46
BUOYS: SMALL (USUALLY 20CM DIAMETER)	67
BUOYS: PIECES (GLASS, PLASTIC, POLYSTYRENE)	66
CRATES (BREAD, BOTTLE)	14

BOTTLE TOPS	74
SEGMENTS OF PLASTIC PIPE	29
PIECES OF ROPE	44
SHOES	25
FLUORESCENT TUBES	6
LIGHT BULBS	6
AEROSOL CANS	7
FOOD/DRINK CANS	7
BIRO TOPS	2

DOLL'S HEADS (1 MALE, 1 FEMALE)	2
COPPER SHEETING FROM HULLS OF WRECKS	8
LORRY TYRE	1
PLASTIC SKITTLE	1
GLUE SYRINGE	1
SMALL GAS CYLINDER	1
CONSTRUCTION-WORKER'S HAT (BROWN)	1
PLASTIC COAT HANGER	1

OCEANS OF GARBAGE

SO LARGE IS THE SEA, SO
VAST AND SO DEEP, THAT WE
HAVE TENDED TO REGARD IT AS
NEARLY INFINITE, CAPABLE
OF ABSORBING ANY REFUSE WE
CHOOSE TO TOSS INTO IT. HOW
WRONG WE ARE. IN EVEN THE
MOST REMOTE CORNERS OF THE
MOST DISTANT OCEANS GARBAGE
IS PILING UP. A FEW YEARS AGO,
EVOLUTIONARY BIOLOGIST TIM
BENTON VISITED THE FARAWAY
PITCAIRN ISLANDS, WHICH,
BY RIGHTS, SHOULD BE ONE OF
THE MOST PRISTINE PLACES
ON EARTH. HE EXPECTED TO
ENCOUNTER A FEW SPECIES
OF PLANTS, ANIMALS, AND
INSECTS. IN THIS LETTER, HE
DESCRIBES WHAT HE FOUND:
A CATALOGUE OF FLOTSAM
AND JETSAM.

The Pitcairn Islands are amongst the most remote in the world (being over 4,500 km from the nearest continental landmass), and consist of four islands: Pitcairn is a small volcanic island, rising some 350 m out of the sea and inhabited by around fifty people descended from the mutineers on HMS *Bounty;* Oeno Atoll (23 56°S, 130 45°W) and Ducie Atoll (24 40°S, 124 47°W) are both uninhabited, low-lying coral atolls (160 and 472 km respectively away from Pitcairn); and Henderson Island (128 22°W, 24 19°S), is a coral atoll raised out of the water to a height of 30 m by the leverage exerted on the seabed due to the weight of the volcano of nearby Pitcairn.

Henderson Island is of enormous scientific interest for several reasons. First it has been uninhabited in modern times, so remains the best Pacific example of an island ecosystem. Second, it is a raised island, so is relatively immune from the cyclonic sea-scour that cleans low-lying islands of colonising plants and animals; as a result colonists have remained and often evolved into island-specific forms. Third, the types of animals and plants to have dispersed the great distances to the island are of biogeographic interest. Fourth, Henderson was once inhabited by a small community of Polynesians (as was Pitcairn), and their untouched remains are exceptionally well-preserved, constituting perhaps the best archaeological site in Polynesia. Fifth, studying Henderson's untouched fossil coral lagoon and its cliffs can give important insights into the processes by which islands, and their biotas, evolve.

In recognition of Henderson's unique importance it was designated a UNESCO World Heritage Site in 1988.

Owing to the remoteness of Henderson, neither it nor the other Pitcairn Islands (both Oeno and, especially, Ducie are important seabird breeding areas) have ever been scientifically well studied. So the Sir Peter Scott Commemorative Expedition to the Pitcairn Islands was planned. This expedition, costing over £100,000, lasted from January 1991 to March 1992 and was divided into three-month phases. Scientists flew to Tahiti where they met a chartered yacht for the two–three week sail to the Pitcairn Islands. Most scientific work was conducted on Henderson, where we lived, but occasional trips were made to the other islands.

My job, on the first phase, was to cut paths (a nightmare) through the thick and tangled forest and, principally, to investigate the insects and spiders that had managed to colonise the islands. Before leaving for Henderson, I knew I would find rubbish on the beaches, but was seriously surprised to find as much as was there. I was on Henderson during the Gulf War, and even though the war was so remote from us we found washed up lots of toy soldiers, toy jeeps, and so on — a constant reminder.

Anyway, we went off to Ducie (three days sail away) for a visit at the end of March 1991. Ducie is periodically scoured by cyclones, and is the remotest island in the Pitcairn group. Resultingly, there is essentially only one plant to be found there, and it didn't take me very long to find all the insects and spiders either. One morning, I was wandering along the beach — being a beach-comber at heart — looking at the garbage, when I thought "the diversity of this stuff is amazing, why not note it down, it will pass the time." By the end of the morning I was hooked — not by the amount of garbage (I recorded 953 objects, which is frightening), but by its diversity — bottles, plastic, toys, beer kegs, pens, medication, food and so on. In total, I transected the seaward edge of the islet of Acadia. Amongst the bizarre items were an English beer keg, a message in a bottle (which, when opened, said "if you find this bottle, please mail to . . ." and the rest was obliterated) and a plastic foot mat from a Peugeot car. On Oeno we even found the pedal from a bicycle, and on Henderson we found a medical syringe and needle. I also found a plastic tube of some cosmetic, though the writing was obliterated by sand-scour. I twisted the top rather roughly and the tube split, squirting me with smelly gunge. We had no fresh water for clothes washing, and the inefficient salt-water washing left the stain, and smell, for weeks afterward!

Early on, we managed to cut a path through the jungle on Henderson to the extreme southern end of the island. The island plateau at the south end is surrounded by 30 m cliffs which drop straight to the sea. Access to this area is only via 7 km paths cut through the thick jungle from the landing places on the north of the island, so the area may have been visited by only a handful of people over the last 1,000 years. Here I also found plastic rubbish — washed up onto the island plateau by the action of storm waves . . .

Comparison of beach surveys on Oeno and Ducie with a beach in SW Ireland reveals that Oeno and Ducie are as dirty as Ireland, and that, barring very ephemeral rubbish such as sweet wrappers, similar sorts of rubbish occur both in the Pacific and Europe in broadly similar proportions. That's kind of scary — places as far away from civilisation as it is almost possible to get on this planet are as dirty as anywhere else. My conclusion from the Pacific work, where bottles occurred on the beaches from a wide-variety of source-countries, was that most litter was illegally dumped from boats.

— TIM BENTON, 1994

DEATH BY BALLOON

EVEN SEEMINGLY INNOCUOUS OBJECTS THOUGHTLESSLY DISCARDED CAN HARM SEA ANIMALS.

THE SPERM WHALE WAS CELEBRATED IN NINE-TEENTH-CENTURY ART AND LITERATURE FOR ITS FIERCE AND NOBLE BEARING (MOBY-DICK WAS A SPERM WHALE). SPERM WHALES ARE NO LONGER ENDANGERED BY HUNT-ING, BUT, LIKE OTHER SEA CREATURES, THEY DO ENCOUNTER ENVIRON-MENTAL HAZARDS. BIO-LOGISTS WHO PERFORMED AN AUTOPSY ON AN EMACIATED MALE SPERM WHALE BEACHED AT SEA SIDE HEIGHTS, NEW JERSEY, IN THE SUMMER OF 1985 FOUND A MYLAR PARTY BALLOON WITH ITS RIBBON STILL ATTACHED (RIGHT) BLOCKING THE ANIMAL'S DIGESTIVE TRACT.

FOR WANT OF FISH TO CATCH

They called it King Cod, and regal it was indeed in economic power and productivity. For more than four hundred years, cod fishing off Newfoundland provided food, jobs, and exports in apparently limitless quantity. And then it all collapsed. Gradually, year by year, sizes of fish decreased and numbers diminished. Changes in water temperature and a decline in the cod's food supply put further pressure on cod populations, until finally, in 1992, the Canadian government had no choice but to temporarily close the Newfoundland cod industry. More than 50,000 people were thrown out of work.

It is probably impossible for anyone now alive to comprehend the magnitude of fish life in the waters of the New World when the European invasion began. It may have been almost equally difficult for the early voyagers. According to the records they have left for us, they seem to have been overwhelmed by the glut of fishes.

In 1497, John Cabot set the tone by describing the Grand Banks as so "swarming with fish [that they] could be taken not only with a net but in baskets let down [and weighted] with a stone." On the lower St. Lawrence in 1535 Jacques Cartier reported that "This river . . . is the richest in every kind of fish that anyone remembers ever having seen or heard of; for from its mouth to its head you will find in their season the majority of the varieties of salt- and fresh-water fish . . . great numbers of mackerel, mullet, sea bass, tunnies, large eels . . . quantities of lampreys and salmon . . . [in the upper River] are many pike, trout, carp, bream and other fresh-water fish." . . .

Early voyagers to the northeastern approaches of America encountered two kinds of land. One was high and dry, and they called it the Main. The other lay submerged beneath 30 to 150 fathoms of green waters, and they called it the Banks. The waters of the continental shelf from Cape Cod to Newfoundland form an aqueous pasture of unparalleled size and fecundity — a three-dimensional one with a volume sufficient to inundate the entire North American continent to a

depth of a yard or more. In 1500, the life forms inhabiting these waters had a sheer mass unmatched anywhere in the world. This was the realm where cod was king.

The name Cabot used for Newfoundland in 1497 was Baccalaos, that being the one bestowed on it by Portuguese who had led the way. The word means, simply, land of cod. And Peter Martyr (from about 1516) tells us that "in the sea adjacent [to Newfoundland, Cabot] found so great a quantity . . . of great fish . . . called baccalaos . . . that at times they even stayed the passage of his ships."

The New World banks, and especially the Grand Banks lying to the eastward of Newfoundland, were a cod fisher's version of the Promised Land. By 1575, more than 300 French, Portuguese, and English fishing vessels were reaping a rich harvest there. Members of Sir Humphrey Gilbert's colonizing venture fairly babbled at the abundance of baccalieu. Cod, wrote one of the visitors, were present "in incredible abundance, whereby great gains grow to them that travel to these parts: the hook is no sooner thrown out but it is eftsoons drawn up with some goodly fish." To which one of his companions added, "We were becalmed a small time during which we laid out hook and lines to take Cod, and drew in, in less than two hours, fish so large and in such abundance that for many days after we fed on no other provision." A third summed it up: "Incredible quantity and no less variety of fishes in

Load of cod near Fogo Island, Newfoundland, July 1990. One of the last good hauls before the fishery closed.

OPPOSITE *Colonizing agents emphasized the amazing natural resources of North America to European readers in the Colonial Era. To promote a favorable view of the New World, John White painted scenes of Indian life in what was to become North Carolina, including one of fishing, in 1585–87. Flemish printer Théodore de Bry added even more marine life to this popular 1590 engraving he made of White's watercolor.*

the seas [especially] Cod, which alone draweth many nations thither and is become the most famous fishing of the world." . . .

At the turn of the sixteenth century, as many as 650 vessels were catching thousands of tons of cod in New World waters, using only baited hooks and handlines. As John Mason, an English fishing skipper working out of a Newfoundland shore station, noted, "Cods are so thick by the shore that we hardly have been able to row a boat through them. I have killed of them with a Pike . . . Three men going to Sea in a boat, with some on shore to dress and dry them, in 30 days will commonly kill between 25 and thirty thousand, worth with the oyle arising from [their livers], 100 to 200 Pounds." . . .

Near the end of the sixteenth century Richard Whitbourne, another fishing skipper, wrote that the average lading for any given ship tallied 125,000 cod. These were from virgin cod populations producing fish up to six or seven feet in length and weighing as much as 200 pounds, in contrast with today's average weight of about six pounds. In Whitbourne's time it was still in the fifteen- to twenty-pound range and the annual cod fishery in the northeastern approaches yielded about 368,000 tons.

By 1620 the cod fleet exceeded 1,000 vessels, many making two voyages annually: a summer one for dry cod and a winter trip from which the cod were carried back to Europe in pickle as "green fish." Yet, despite the enormous destruction, there was no apparent indication that cod stocks were diminishing. As the seventeenth century neared its end, travellers such as Baron Lahontan were still writing about the cod as if its population had no bounds.

"You can scarce imagine what quantities of Cod-fish were catched by our Seamen in the space of a quarter of an hour . . . the Hook was no sooner at the bottom than a Fish was catched . . . [the men] had nothing to do but to throw in, and take up without interruption . . . However, as we

were so plentifully entertained at the cost of these Fishes, so such of them as continued in the Sea made sufficient reprisals upon the Corps of a Captain and several Soldiers, who dy'd of the Scurvy, and were thrown over-board."

The first hint that the destruction might be excessive (and it is a veiled hint) comes from Charlevoix in the 1720s. After first telling us that "the number of the cod seems to equal that of the grains of sand," he adds that "For more than two centuries there have been loaded with them [at the Grand Banks] from two to three hundred [French] ships annually, notwithstanding [which] the diminution is not perceivable. It might not, however, be amiss to discontinue the fishery from time to time [on the Grand Bank], the more so as the gulph of St. Lawrence [together with] the River for more than sixty leagues, the coasts of Acadia, those of . . . Cape Breton and of Newfoundland, are no less replenished with this fish than the great bank. These are true mines, which are more valuable, and require much less expense than those of Peru and Mexico." That Charlevoix was not exaggerating the value of the cod fishery is confirmed by the fact that, in 1747, 564 French vessels manned by 27,500 fishermen brought home codfish worth a million pounds sterling — a gigantic sum for those days. . . .

By 1800, English- and French-based vessels had become notably fewer, but Newfoundlanders, Canadians, and Americans more than made up the loss. In 1812, 1,600 fishing vessels, largely American, were in the Gulf, with as many more Newfoundland and Nova Scotia ships fishing the outer banks and the Atlantic coast of Labrador.

Those were the days of the great fleets of "white wings," when the sails of fishing schooners seemed to stretch from horizon to horizon. In addition to this vessel fishery, thousands of inshore men fished cod in small boats from every little cove and harbour. Vesselmen and shoremen alike still mostly fished in the old way with hooks and lines because "the glut of cod" was still so great

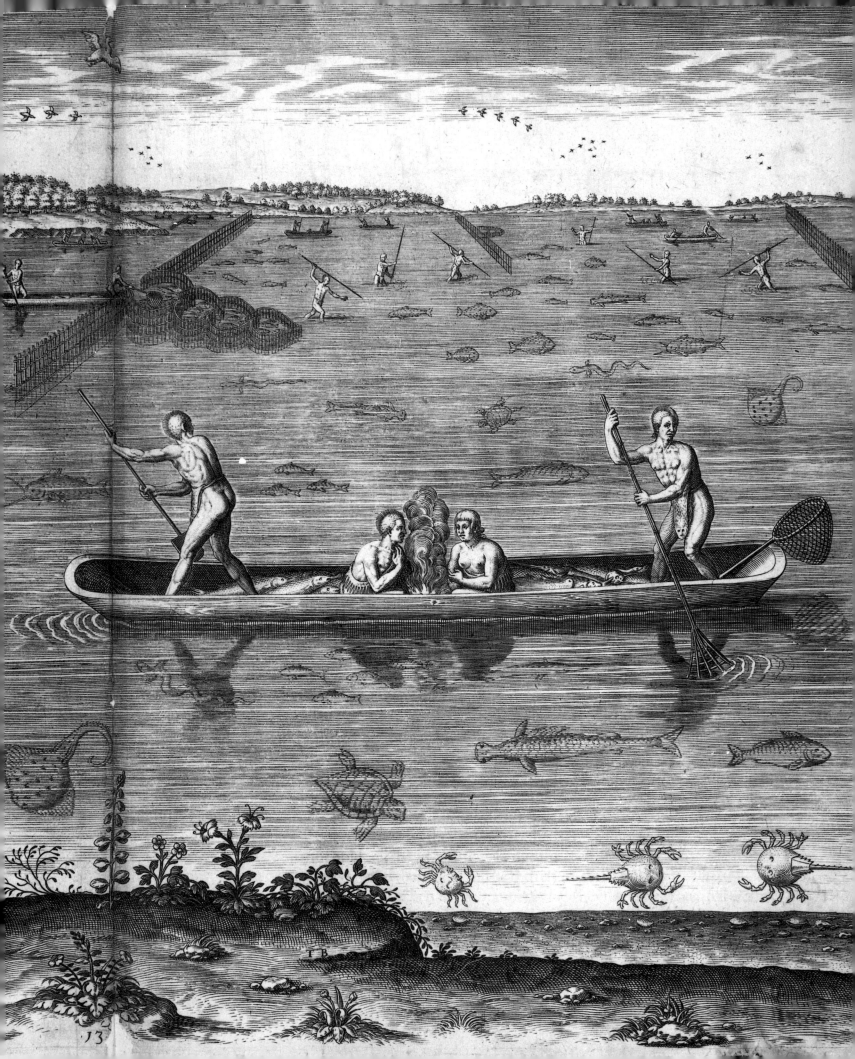

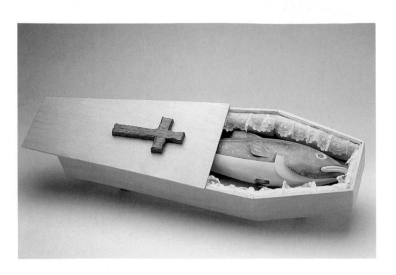

ABOVE *Cod coffin, carved wood, 1994. Fisherman Dan Murphy made this symbol of the closing of the cod fishery in 1992. Newfoundland fishermen have few options for alternative employment.*

RIGHT *Atlantic cod inside a fishing net, Lofoten Islands, Norway. Cod are bottom dwellers and voracious feeders on mollusks, worms, crabs, and many species of fish.*

LEFT *Landing a dory loaded with cod from nets at sea, Bonne Esperance, Canada, circa 1930. Among the many commercial uses of cod was the dietary supplement cod-liver oil, the scourge of generations of children before World War II.*

that nothing more sophisticated was needed.

In 1876, John Rowan went aboard "a schooner cod-fishing close to shore. . . . They were fishing in about three fathoms of water and we could see the bottom actually paved with codfish. I caught a dozen in about fifteen minutes; my next neighbour [a crewman] on the deck of the schooner, caught three times as many, grumbling all the time that it was the worst fishing season he had ever known."

Between 1899 and 1904, the annual catch of cod (and of haddock, which in the salt-fish business was treated as cod) approached a million tons. During those years, Newfoundland alone annually exported about 1,200,000 quintals of dry fish, representing about 400,000 tons of cod, live weight. By 1907, the Newfoundland catch had risen to nearly 430,000 tons; and there were then some 1,600 vessels, of many nationalities, fishing the Grand Banks.

But now there was a chill over the Banks — one that did not come from the almost perpetual fog. Cod were getting harder to catch, and every year it seemed to take a little longer to make up a voyage. At this juncture, nobody so much as breathed the possibility that the Banks were being over-fished. Instead, one of the age-old fisherman's explanations for a shortage was invoked: the cod had changed their ways and, temporarily, one hoped, gone somewhere else.

The early nineteenth-century discovery of immense schools of cod along the Labrador coast even as far north as Cape Chidley was seen as confirmation that the fish had indeed changed their grounds. In actual fact the Labrador cod comprised a distinct and, till then, virgin population. They did not stay that way for long. By 1845, 200 Newfoundland vessels were fishing "down north" and by 1880, up to 1,200. As many of 30,000 Newfoundlanders ("floaters" if they

fished from anchored vessels, and "stationers" if they fished from shore bases) in 1880 were making almost 400,000 quintals of salt cod on the Labrador coast alone.

The Labrador cod soon went the way of all flesh. The catch steadily declined thereafter until, by mid-twentieth century, the once far-famed Labrador fishery collapsed. Attempts were again made to ascribe the disappearance of the Labrador fish to one of those mythical migrations. This time it did not wash. The fact was that King Cod was becoming scarce throughout the whole of his wide North Atlantic realm. In 1956, cod landings for Grand Banks/Newfoundland waters were down to 80,000 metric tonnes — about a fifth of what they had been only half-a-century earlier.

When a prey animal becomes scarce in nature, its predators normally decrease in numbers, too, permitting the prey an opportunity for recovery. Industrial man works in the opposite way. As cod became scarcer, so did pressure on the remaining stocks mount. New, bigger, more destructive ships came into service and the bottom trawl, which scours the bed of the ocean like a gigantic harrow, destroying spawn and other life, almost totally replaced older fishing methods. Scarcity brought ever-rising prices, which in turn attracted more and more fishermen. During the 1960s, fleets of big draggers and factory ships were coming to the Banks from a dozen European and Asian countries to engage in a killing frenzy over what remained of the cod populations. The result was that, between 1962 and 1967, cod landings increased until, in 1968, the catch topped two million tons. Soon thereafter, the whole northwest Atlantic cod fishery disintegrated for want of fish to catch.

— FARLEY MOWAT, *Sea of Slaughter,* 1984

HEAVY FISHING

THE UNITED NATIONS FOOD AND AGRICUL-
TURE ORGANIZATION ESTIMATES THAT OF THE
SEVENTEEN MAJOR FISHERIES AREAS IN THE
WORLD, FOUR ARE DEPLETED AND THE OTHER
THIRTEEN ARE EITHER FISHED TO CAPACITY
OR OVERFISHED.

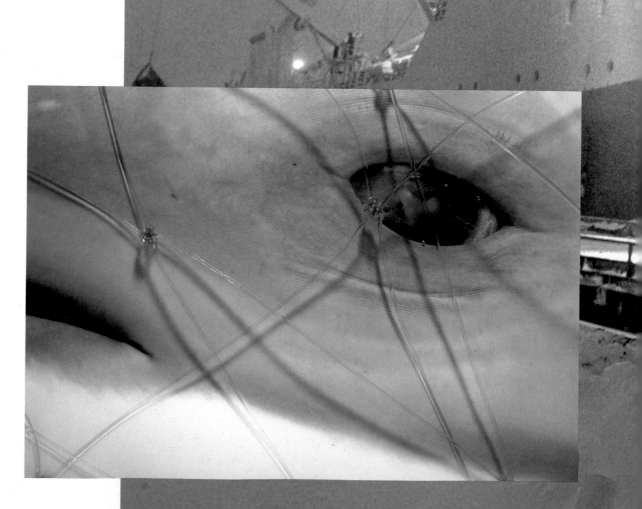

RIGHT WHITE-SIDED
DOLPHIN DROWNED IN
DRIFTNET, NORTH
PACIFIC, 1990. BEFORE
DRIFTNETTING ON THE
HIGH SEAS WAS BANNED
BY THE UNITED NATIONS
IN 1993, UP TO TEN
THOUSAND DOLPHINS AND
WHALES AND MILLIONS
OF SHARKS WERE KILLED
IN DRIFTNETS EVERY
YEAR.

OPPOSITE F/T *SAGA
SEA* UNLOADING POLLOCK
AND FISH MEAL DURING
A BLIZZARD IN DUTCH
HARBOR, UNALASKA,
ALASKA, 1992. LARGE
FACTORY TRAWLERS
CAN PROCESS 550 TONS
OF FISH A DAY.

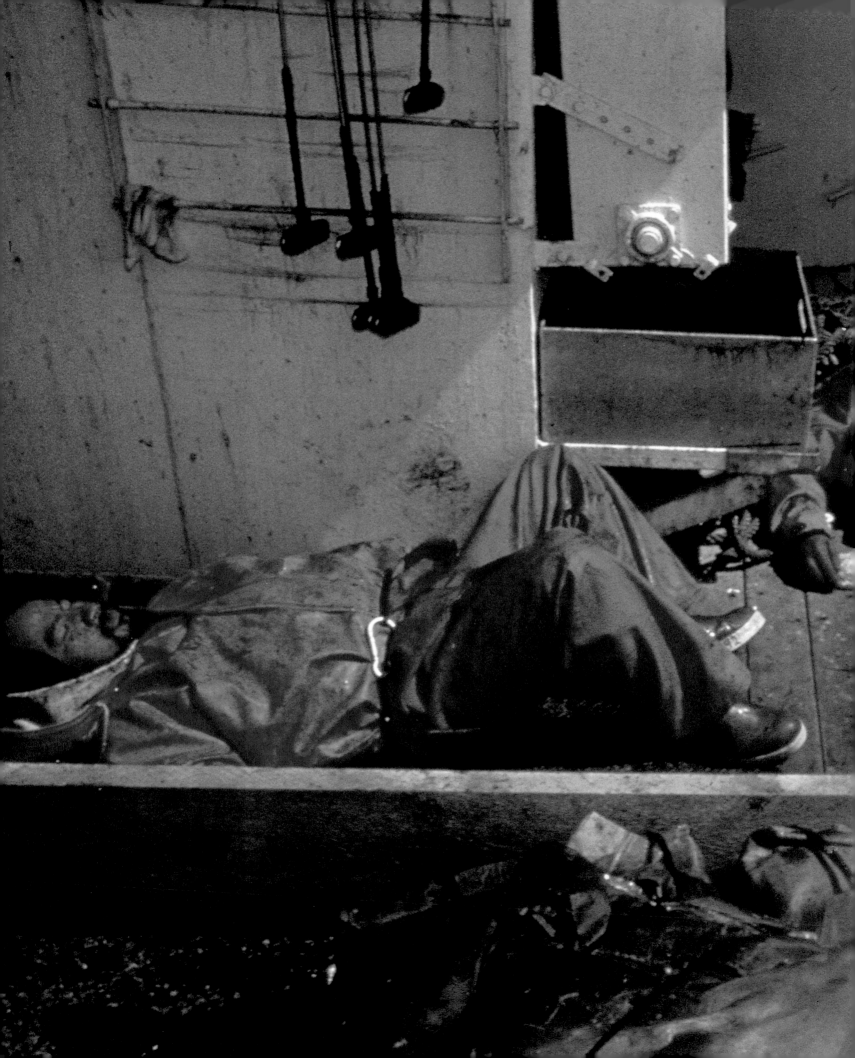

EXHAUSTED FISHERMEN
COLLAPSE AT THE END
OF THE 1992 "HALIBUT
DERBY." ONE WAY TO
TRY TO CONTROL FISHING
PRESSURE ON DWINDLING
RESOURCES IS TO LIMIT
THE TIME DURING WHICH
FISH CAN BE CAUGHT.
BY 1992 THE ENTIRE
UNITED STATES COMMER-
CIAL FISHING SEASON
FOR PACIFIC HALIBUT
OFF ALASKA HAD BEEN
PARED TO TWO TWENTY-
FOUR-HOUR PERIODS PER
YEAR. FOR TWO DAYS,
FISHERMEN WORKED
ROUND THE CLOCK, RISK-
ING LIFE AND LIMB TO
CATCH ABOUT 60 MILLION
POUNDS OF HALIBUT.

TRASH FISH

THROUGHOUT THE WORLD, THE MOST VALU-
ABLE SPECIES OF FISH ARE HEAVILY FISHED
OR DEPLETED. FISHERMEN FIND THAT THEIR
NETS ARE FULL OF LESS VALUABLE "TRASH
FISH." MANY OF THESE SPECIES ARE ACTU-
ALLY QUITE TASTY, AND FISHERMEN AND
FISHERMEN'S WIVES' ORGANIZATIONS WORK
TO CREATE MARKETS FOR THESE UNPOPULAR
SPECIES. THE RECIPE PRESENTED HERE
REPRESENTS AN EFFORT TO MARKET THE
SHARK-RELATIVES THAT CROWD THE NETS
OF COD FISHERMEN.

STIR-FRY SKATE WITH VEGETABLES

2 POUNDS SKATE WING (FILLETED)
FLOUR
3 TABLESPOONS BUTTER OR OIL
2 CLOVES GARLIC, FINELY CHOPPED
2-3 FRESH SPRIGS OF PARSLEY
JUICE OF 2 LEMONS
1/2 CUP WHITE WINE (MORE IF NEEDED)
VEGETABLES: CARROTS, ASPARAGUS, ZUCCHINI,
    RED PEPPER, SNOW PEAS, ETC.

*Cut skate into long thin strips, using the cartilage indentations
as a guide. Fillets 8 oz. or larger work best for this recipe
(although the cartilage has been removed by filleting, you can
see where they were.) Using a wok or heavy skillet, saute garlic
briefly in butter or oil. Add parsley and vegetables and stir-fry
for one minute; add skate (skate should be dredged in flour).
Stir for 3-4 minutes, or until skate is cooked through. Deglaze
with wine and stir for 15 more seconds. This recipe will go
over rice or noodles very nicely. Serves 4 or more depending on
how much pasta or rice is used.*

— From The Taste of Gloucester: *A Fisherman's Wife Cooks,
written and compiled by the Fishermen's Wives of Gloucester
and the Cape Ann League of Women Voters, 1976*

A HAUL OF SKATES
INSTEAD OF COD,
GEORGES BANK OFF
NEW ENGLAND, 1985.

HABITAT IS WHERE IT'S AT

When we flush a toilet or chop down a tree or build a parking lot, we may believe that each is an isolated act with no consequences beyond itself. We're fooling ourselves. Nearly everything we do on or near rivers, streams, or coastal waters affects the ocean. Logging around salmon streams, for example, alters breeding conditions; cutting trees in tropical forests causes silt runoff that chokes coral reefs; sewage generates phosphates and nitrates that upset the ecological balance of coastal waters; and uncontrolled development decimates habitats. Everything needs a place to live, to feed, to grow, to spawn. If we alter or destroy the living space, we jeopardize the living things.

OPPOSITE *Sockeye salmon swimming upstream to spawn, Brooks River, Katmai National Park, Alaska.*

RIGHT *Logging an old-growth sitka spruce, Lyell Island, British Columbia. Logging old-growth trees may cost more jobs in the salmon industry than the number of jobs that are at risk from restrictions on cutting.*

When you think about how hard it is to get to be a steelhead in the Columbia River, it's a wonder any of them make it. First, you're a fertilized egg in a pocket of gravel on the bottom of a tributary stream somewhere, and if you're lucky the timber is still standing on the banks and nearby hills. If it's gone to make gum wrappers, cardboard boxes, or subdivisions, the stream will fill up with silt from erosion and that's all she wrote for you and the rest of the salmon and steelhead eggs. And when the parent fish get back the next year, they're not going to find any place to cuddle, either.

The next big test in your steelhead childhood comes after your egg develops an eye and you turn into an actual little fish with fins and everything. Birds and other fish already will have eaten a bunch of your cousins, and that's going to keep on, like it or not. If you last until you're about two or three inches long, though, you have to start thinking about the turbines in the dams downstream. You're an anadromous fish, and that means you head out to the Big O to spend most of your life because that's where the living is easy — salt water that doesn't freeze, lots of food, no bears, eagles, and like that. The trouble is that between you and the ocean are about ten dams, each rigged with turbines that churn like giant Cuisinarts making salmon and steelhead puree.

Always, you have to worry about not getting eaten, and about where your next meal grew up itself. The closer you get to the coast, the more birds, seals, sea lions, and bigger fish you have to deal with, and the more trouble you can get into eating a herring or needlefish that's been hanging around a pulp mill or sewer outfall someplace sucking up PCBs or petroleum sludge. Finally, if you survive all that, you cruise for a couple or three years out in the ocean, grow up, then head back through the same mess to try to score in the stream of your birth — if it's still there.

Steelhead share stream and nearshore habitat in common with about 90 percent of the fish we depend on for food and sport. In the relative shallows of the coastal zone, light penetrates to invigorate the water column from top to bottom,

177

producing a wonderful chow line of plant and animal plankton, and every kind of fish from there up. It's a great place for a fish to raise a kid, and that's why beaches, bays, and estuaries are so critical to keeping the ocean food web intact.

The problem is that waterfront occupation by the booming human population destroys the rearing areas for several stages in the food web. When you trash the sardines in southern California by either catching all of them or building Los Angeles, you also kill off the marlin, swordfish, tuna, and on and on and on. It's not hard to figure out what to do: don't dump poison in the ocean and don't destroy estuaries by filling them in to build shopping malls.

For a long time we thought the ocean was a great garbage disposal, and for things that disintegrate it still is. Plastic, however, and other engineered substances last hundreds, even thousands of years and you only have to visit one beach near a city to know what that means. About 85 percent of the pollution of the ocean comes from what we do on land, and we also now know that almost all of that never leaves the coast. Getting to be a steelhead is harder and harder every year.

— BRAD MATSEN,
Ray Troll's Shocking Fish Tales, 1991

A PLACE TO LIVE

VERDANT FORESTS, GRASSY PLAINS, ROCKY
CLIFFS, AND COLORFUL TROPICAL LAND-
SCAPES ARE HIDDEN UNDER THE SEA.
EVERY CREATURE'S HABITAT IS MAINTAINED
BY A COMPLEX EQUILIBRIUM OF DIFFERENT
FACTORS, SUCH AS WATER TEMPERATURE,
WATER CHEMISTRY, LIGHT, AND BOTTOM
CHARACTERISTICS. WHEN ONE OR MORE OF
THESE CONDITIONS ARE ALTERED, THE
ENTIRE ECOSYSTEM MAY BEGIN TO DECLINE.
COASTAL ECOSYSTEMS THROUGHOUT THE
WORLD ARE THE FIRST TO FEEL THE EFFECT
OF HUMAN ACTIVITIES.

CORAL REEF, RED
SEA. CORAL REEFS ARE
DAMAGED BY HUMAN
ACTIVITY ON LAND AND
ALONG THE COASTS.
SILT FROM LOGGING AND
OVER-FISHING ALSO
POSE MAJOR PROBLEMS.
WHEREVER REEFS ARE
CLOSE TO GROWING
COASTAL POPULATIONS,
THEY ARE IN SEVERE
DECLINE.

LEFT KELP FOREST,
CALIFORNIA. KELP
FORESTS SHELTER MANY
KINDS OF LIFE IN THE
TEMPERATE COASTAL
WATERS OF THE
AMERICAS, EUROPE,
AND ASIA, BUT THESE
HABITATS ARE VULNERA-
BLE TO DECLINING
WATER QUALITY, OVER-
HARVESTING, AND
FLUCTUATIONS IN WATER
TEMPERATURE.

ABOVE MANGROVE
ROOTS, PELICAN CAYS,
BELIZE. MANGROVES
GROWING ALONG THE
MUDDY MARGINS OF TROP-
ICAL OCEANS ARE NURS-
ERY AREAS FOR YOUNG
FISH AND SHELLFISH,
AND THEY CONTROL
EROSION AND WATER
QUALITY. NEARLY HALF
OF THE WORLD'S MAN-
GROVE FORESTS AND
SALT MARSHES HAVE
BEEN CLEARED, DRAINED,
DIKED, OR FILLED.

COLONY OF GENTOO
PENGUINS, ANTARCTICA.
EVEN DISTANT WATERS
OF THE ANTARCTIC ARE
SUBJECT TO POLLUTION
AND OTHER FORMS OF
HUMAN DISTURBANCE.
RECENTLY, TRACE
AMOUNTS OF DDT WERE
FOUND IN THE EGGS OF
ANTARCTIC PENGUINS.

HELP FOR OCEAN PLANET

The dangers to the Ocean Planet are dire, but it's not too late. Taking small measures in our daily routines can vastly improve the oceans' outlook.

WHAT YOU CAN DO . . .

To control runoff and erosion

➤ drive less, walk, bike, and use car pools or public transportation

➤ recycle used motor oil and antifreeze

➤ re-sod grass or replace with native plants and divert water from pavement onto grass

To minimize nutrient build-up

➤ choose the right fertilizer and use it conservatively

➤ don't overwater

➤ use non-phosphate detergents

To reduce marine debris

➤ recycle glass, aluminum, plastic, and paper

➤ precycle: be aware of packaging, especially plastics

➤ buy recycled products

➤ never release balloons outside

➤ break or cut all six-pack rings

➤ if you bring plastic on a boat or to a beach, dispose of it properly

To take care with toxic chemicals

➤ use fewer pesticides; store and dispose of properly

➤ drain swimming pools and dispose of treatment chemicals properly

➤ find out about and use battery and oil recycling programs and household hazardous waste collection programs

➤ learn about and use safe alternatives to household hazardous chemicals

➤ use rechargeable batteries

To conserve water and energy

➤ take shorter showers; install efficient shower heads, faucet aerators, and toilets; fix leaky toilets and faucets

➤ turn thermostats down; and turn off appliances and lights when not in use

➤ run dishwashers and washing machines only with full loads; insulate hot-water heaters

➤ learn about xeriscaping — landscaping to conserve water use

To take action

➤ adopt and clean up a beach or stream

➤ write to your local policy makers and newspaper editors

➤ join a local citizens' environmental program

➤ teach your children, parents, and friends

WHAT OTHERS HAVE DONE . . .

➤ Wilson Vailoces, a sixty-year-old farmer from the Bindoy Province in the Philippines, started Sea Watchers, a volunteer organization to patrol for illegal destructive fishing practices and replant lost mangroves. He has planted nearly 10,000 mangrove trees himself.

➤ A French and British ban on boat paint containing tributyl tin produced almost immediate improvement in poisoned oyster populations.

➤ Each year 160,000 volunteers from 33 countries participate in the International Coastal Clean-up.

➤ In November, 1994, the United Nations Law of the Sea treaty took effect. This "constitution for the oceans" governs seabed mining, environmental protection, exploitation of natural resources, as well as jurisdiction, access to the seas, scientific research, and settlement of disputes.

➤ Reef Relief, a Florida group working to prevent anchor damage to coral reefs, has installed more than one hundred mooring buoys in Key West.

➤ The United States government National Marine Sanctuary Program protects and manages more than 18,500 square miles of coastal and ocean waters.

➤ Concern for the health of the Antarctic ecosystem prompted world leaders to negotiate CCAMLR (Convention for the Conservation of Antarctic Marine Living Resources). CCAMLR regulates all fishing activities in the Southern Ocean and requires that fisheries management consider potential effects on the entire ecosystem, instead of focusing solely on the fish.

➤ The International Marine Life Alliance trains Philippine fishermen to use nets instead of harmful cyanide to catch aquarium fish, and campaigns around the world against cyanide-caught aquarium fish.

➤ Turtle excluder devices, now required on shrimp nets throughout the Gulf of Mexico, Caribbean, and western Atlantic, reduce sea turtle kill by 97 percent and still capture 98 percent of the shrimp that enter the nets.

➤ Compact fluorescent light bulbs use only one-fourth as much electricity as conventional incandescent bulbs. Although socket-type fluorescent bulbs are more expensive to buy, they last about thirteen-times longer and save about three times what they cost over the life of the bulb.

ABOUT THE CONTRIBUTORS

WILLIAM BEEBE (1877–1962) wrote popular natural history books, including *Jungle Days* and *Half Mile Down.* A naturalist and ornithologist, he served for more than thirty-five years as director of the Department of Tropical Research for the New York Zoological Society.

TIM BENTON is an editor at Cambridge University Press. He is an evolutionary biologist specializing in the evolution of behavior and life-histories of invertebrate animals.

RACHEL CARSON (1907–1964) was the author of several influential books, including *The Sea Around Us,* which won the National Book Award, and *Silent Spring,* which helped to encourage laws that protect the environment. She received many honors for her writing and environmental activism.

JOSEPH CONE is the author of *Fire Under the Sea* and *A Common Fate: Endangered Salmon and the People of the Pacific Northwest.* He is the communications director of Oregon Sea Grant, a marine research and education program.

JOSEPH CONRAD (1857–1924) wrote many powerful books, including the novels *Lord Jim* and *Heart of Darkness.* For sixteen years, he was a seaman in the British merchant navy, and much of his writing drew upon his life at sea.

JACQUES-YVES COUSTEAU is the author of a number of popular books about the oceans, including *The Silent World* and *The Living Sea.* Since 1957, he has been the director of the Oceanographic Museum and Institute in Monaco. His popular television shows and films have created significant public interest in the oceans and their preservation.

ANN DAVISON became the first woman to sail across an ocean alone in 1953. Her book, *My Ship is So Small,* based on her journal, told the story of her voyage.

JAN DEBLIEU is the author of *Hatteras Journal* and *Meant to Be Wild: The Struggle to Save Endangered Species Through Captive Breeding.* She is an essayist and journalist who lives on the Outer Banks of North Carolina

LOREN EISELEY (1907–1977) was the author of *Darwin's Century, The Immense Journey,* and other books. An anthropologist, he wrote eloquently about a wide variety of scientific subjects.

BENJAMIN FRANKLIN (1706–1790) was a printer, author, philosopher, diplomat, scientist, and inventor. His books, *Poor Richard's Almanack* and *Autobiography,* are still widely read. The discovery of the Gulf Stream was his contribution to oceanography.

ASAE FUKUDA testified at an international forum on Minamata disease in 1988. That year the former president and plant manager of the Chisso Corporation were found to be criminally liable for the pollution in Minamata Bay and long-standing issues of compensation to the victims were finally resolved.

KENNETH GRAHAME (1859–1932) published four books, including the classic *The Wind in the Willows,* which he wrote for his son while he was a secretary of the Bank of England.

JAMES HAMILTON-PATERSON is the author of *The Great Deep* and *Playing with Water.* In addition to being a journalist, he is a poet and short-story writer. He is a member of the Royal Geographic Society and lives in Italy and the Far East.

TOM HORTON is the author of *Bay Country,* which won the John Burroughs Medal for natural history writing. He was born and raised on the Eastern Shore of Maryland, and worked as a reporter for the *Baltimore Sun* in 1972–87.

BRAD MATSEN is the author of *Northwest Coast, Deep Sea Fishing,* and other books and articles about fishing and the evolution of life. He has spent most of his life in Alaska and the Pacific Northwest and has worked as a charter pilot, commercial fisherman, and merchant seaman.

PETER MATTHIESSEN is the author of six novels, including *At Play in the Fields of the Lord, Far Tortuga,* and *Killing Mister Watson,* and several books of nonfiction.

JOHN MCPHEE is the author of many books, including *Coming into the Country, Basin and Range,* and *Assembling California.* He has written about a broad variety of subjects, from geology to art.

FARLEY MOWAT is the author of a number of popular books, including *Never Cry Wolf* and *Woman in the Mist.* Much of his writing focuses on the Arctic and injustice toward native peoples and wildlife.

JOHN P. WILEY, JR., is the author of *Natural High* and a columnist and editor of *Smithsonian Magazine.* He is an avid naturalist.

CHARLES ZERNER, director of the Natural Resources and Rights Program of the Rain Forest Alliance, is a Southeast Asia specialist and a lawyer with experience in the analysis of community-environment relationships, property rights, culture, and conservation issues.

ACKNOWLEDGMENTS

The Ocean Planet project received generous funding from the National Science Foundation, The Pew Charitable Trusts, Smithsonian Institution, Geraldine R. Dodge Foundation, National Ocean Industries Association, The David and Lucile Packard Foundation, and The Rockefeller Foundation.

I would also like to thank Times Mirror Magazines, Inc., Ocean Planet's National Corporate Sponsor.

Additional valuable support for the Ocean Planet project was provided by the National Aeronautics and Space Administration, Joint Oceanographic Institutions, Inc., Harbor Branch Oceanographic Institution, Inc., National Oceanic and Atmospheric Administration, National Marine Fisheries Service-NOAA, the Stroud Foundation, Swim Environmental Awareness, W. Alton Jones Foundation, Boston Whaler, National Fish and Wildlife Foundation, and Mme. Tomo Kikuchi.

It was more than a little daunting to assemble an anthology about a subject as vast as the oceans themselves. Many people helped craft this book by suggesting literature sources, providing background material, and checking facts. I am particularly grateful to Peter Benchley for the depth of his knowledge about the oceans and his evocative writing, and to Eric Himmel, our editor at Harry N. Abrams, Inc., who shared fully in the tasks involved in bringing this book to completion.

A tremendous amount of effort went into developing the original material in the Ocean Planet exhibition and this book. We were fortunate to have a talented researcher in Lucinda Leach, who also assembled many of the fact lists. Many of the captions come from the Ocean Planet exhibition writer/editor, Sue Voss. I am also indebted to Ocean Planet program staff who worked on exhibition research and helped with this book: Beth Nalker, Jodi Asarch, Doyle Rice, Manoj Shivlani, Susan Boa, Karen Roach, Debra Goldstein, Kate Roosevelt, Jon Kohl, Elizabeth Sheehan, Karen Lee, Ann Burrola, Joe Madeira, Lisa Tamaro, Cissy Anklam, and Marjory Stoller. Diane L. Nordeck and Doc Dougherty photographed all of the exhibition objects.

Ocean Planet exhibition advisers helped develop exhibit concepts, reviewed drafts of scripts, and generously gave time and expertise to the project: Fred Abatemarco, William C. Baker, Elisabeth Mann Borgese, Minda Borun, David Cottingham, Judith Diamond, John W. Farrington, Gene Feldman, Kathy Fletcher, J. Frederick Grassle, Jack Horstmeyer, John A. Knauss, Bonnie J. McCay, Laura McKie, Roger McManus, James V. O'Connor, David L. Pawson, Mary Rice, David B. Rockland, Clyde Roper, Carl Safina, Kathryn Sullivan, John R. Twiss, James D. Watkins, Michael Weber, and George Woodwell. More than one hundred and fifteen other exhibition reviewers also contributed to the veracity and timeliness of Ocean Planet content.

At several key junctures Smithsonian staff provided support that made the Ocean Planet project possible. Robert McC. Adams, Constance Newman, Robert Hoffmann, Thomas Lovejoy, Frank Talbot, T. C. Benson, Robert Sullivan, Stanwyn Shetler, Marsha Shaines, Nancy Fischer, Marie Mattson, Randall Kremer, and Francine Berkowitz have all been lifesavers.

— *Judith Gradwohl*

Times Mirror Magazines, Inc. wishes to thank Ford Motor Company, Motorola, Inc., The Discovery Channel and The Water Foundation Radio Network for their generous support of the Ocean Planet education program, of which this book is an important part.

CREDITS

p. 116 (top): © Mary Jane Adams, Arcadia, California

p. 116 (bottom): © David Wrobel, Monterey, California

p. 117 (top): Animals Animals © 1994 K. Atkinson/OSF

p. 117 (bottom): © Mike Severns/ Tom Stack & Associates, Colorado Springs

pp. 118-119: Courtesy National Museum of Natural History, photograph by Chip Clark

p. 120: Courtesy National Museum of Natural History, photograph by Diane L. Nordeck

p. 123: © Kay Chernush, Arlington, Virginia

pp. 124-25: © Tom Campbell, Santa Barbara

p. 126: © Tom Campbell, Santa Barbara

p. 127: © Tom Campbell, Santa Barbara

pp. 128-29: © Doc White/Images Unlimited, Inc.

p. 130: Dann Blackwood/U.S. Geological Survey

pp. 132-33 (top): From *Science*, 264 (1964), pp. 46-7. © AAAS

p. 132 (bottom): Ken C. Macdonald, Dan Scheirer, and Donald Forsyth, University of California at Santa Barbara

p. 133 (bottom): Ken C. Macdonald, Dan Scheirer, and Donald Forsyth, University of California at Santa Barbara

pp. 136-37: © Michael Baytoff, Flemington, New Jersey

p. 140: W. Eugene Smith/Black Star, New York. © Aileen M. Smith

p. 143: W. Eugene Smith/Black Star, New York. © 1981 The Heirs of W. Eugene Smith

pp. 144-45: Frank Hewetson/ © Greenpeace

pp. 146-47: © Natalie Fobes, Seattle

p. 148: © David W. Harp, Baltimore

pp. 150-51: © Robert W. Madden/National Geographic Society

p. 153: © David H. Harvey/Woodfin Camp & Associates, New York

p. 154: © Robert Perron, Banford, Connecticut

p. 155: © Robert Perron, Banford, Connecticut

p. 156: Tim Benton, Cambridge, England

p. 159: Charles Fowler, National Marine Mammal Laboratory, Seattle, Washington

p. 160: Courtesy Marine Mammal Stranding Center, photograph by Diane L. Nordeck

pp. 160-161: William Heysman Overend. *Harpooning a Whale*. c. 1870. Oil on canvas, 19 3/8 x 29". The Mendall Whaling Museum, Sharon, Massachusetts

p. 162: © Gordon Peterson, Pincher Creek, Canada

p. 165: © The British Library, London

p. 166: Carving by Dan Murphy, photograph by Diane L. Nordeck

pp. 166-67: © Doug Allan/Oxford Scientific Films, England

p. 169: George Whiteley

p. 170: R.O. Grace/© Greenpeace

pp. 170-71: © Natalie Fobes, Seattle

pp. 172-73: © Bob Sacha, New York

p. 175: © Richard Howard, Winchester, Massachusetts

p. 176: © Jeff Foott, Jackson, Wyoming

p. 177: © Steve Jackson/URSUS Photography, Vancouver, Canada

p. 178: Earth Scenes © 1994 Doug Wechsler

p. 179: © Gary Baarsch/Woodfin Camp & Associates, New York

pp. 180-81: © Jeffrey L. Rotman, Somerville, Massachusetts

pp. 182-83: © Norbert Wu, Orinda, California

p. 183: © Tony Rath, Belize

p. 184: © Art Wolfe, Seattle

p. 185: © Onne Van Der Wal/Stock Newport, Rhode Island

pp. 186-87: © Helmut Horn/Alextra

TEXT CREDITS

Excerpt from *The Edge of the Sea* by Rachel Carson. Copyright © 1955 by Rachel L. Carson, © renewed 1983 by Roger Christie. Reprinted by permission of Houghton Mifflin Company. All rights reserved.

Excerpts from "The Star Thrower" in *The Unexpected Universe* by Loren Eiseley. Copyright © 1969 by Loren Eiseley, reprinted by permission of Harcourt Brace & Company.

"Beachwalker" reprinted with permission from "Phenomena, Comment and Notes" by John P. Wiley, Jr., *Smithsonian*, July 1985.

Excerpt from *My Ship Is So Small* by Ann Davison. Copyright © 1956 Peter Davies, London.

Excerpt from *Men's Lives* by Peter Matthiessen. Copyright © 1986 by Peter Matthiessen. Reprinted by permission of Random House, Inc.

Excerpts from *Looking for a Ship* by John McPhee. Copyright © 1990 by John McPhee. Reprinted by permission of Farrar, Straus & Giroux, Inc.

Excerpt from *Half Mile Down* by William Beebe. Copyright © 1934 Harcourt, Brace & Company.

Excerpt from *The Silent World* by Jacques-Yves Cousteau. Copyright © Jacques-Yves Cousteau. Reprinted by permission of the author.

Excerpt from *Fire Under the Sea* by Joseph Cone. Copyright © 1991 Joseph Cone. By permission of William Morrow & Company, Inc.

Excerpt from *The Great Deep* by James Hamilton-Paterson. Copyright © 1992 by James Hamilton-Paterson. Reprinted by permission of Random House, Inc.

Testimony by Asae Fukuda reprinted by permission of Keiso Shobo Publishing Company, Tokyo.

Excerpt from "Chesapeake Bay — Hanging in the Balance" by Tom Horton reprinted by permission of the *National Geographic Society*.

Excerpt from *Sea of Slaughter* by Farley Mowat. Copyright © 1984 by Farley Mowat Ltd. By permission of Little, Brown and Company.

"Habitat Is Where It's At" excerpted from *Shocking Fish Tales* by Ray Troll. Copyright © 1991 Ray Troll. Used by permission of Celestial Arts, P.O. Box 7327, Berkeley, CA 94707.

Ocean Planet is a major traveling exhibition from the Smithsonian Institution that promotes the celebration, understanding, and conservation of the world's oceans. Premiering at the National Museum of Natural History, Washington, D.C., and touring to eleven U.S. cities, Ocean Planet shows how all of our lives are connected to the oceans.

On the cover: Breaking wave, Waimea Bay, Oahu, Hawaii. Photograph by Warren Bolster © Tony Stone Worldwide

Project manager (for Harry N. Abrams, Inc.): Eric Himmel
Design: Beth A. Crowell, Cheung/Crowell Design
Research: Lucinda Leach
Smithsonian Institution research and staff support:
Susan Boa, Doyle Rice, Manoj Shivlani
Exhibition caption material: Sue Voss
Photographic research: Jodi Asarch, Lucinda Leach,
Cynthia Van Roden
Picture rights: Ann Burrola

Library of Congress Cataloging-in-Publication Data

Benchley, Peter
    Ocean planet / original text by Peter Benchley;
    edited by Judith Gradwohl.
    p. cm.
    Companion volume to a traveling exhibition.
    ISBN 0-8109-3677-1 (Abrams: cloth)
    ISBN 0-8109-2604-0 (Museum: pbk.)
    1. Marine Biology. 2. Ocean. 3. Fisheries. 4. Seafaring life.
5. Marine pollution. 6. Marine resources conservation.
I. Gradwohl, Judith. II. Title.
QH91 .B415 1995
333.95'2 — dc20   94-23265

Printed and bound in Japan